# Site Surveying Guide for Radar and Cameras

A How-To Guide for the Security Professional

Eddie R Hughes

Deep Sea Publishing, LLC

Copyright © 2011 Eddie R Hughes

All rights reserved. Published in the United States by:

Deep Sea Publishing LLC, Herndon, Virginia.

**ISBN-13: 978-0983427612**
**ISBN-10: 0983427615**

Printed in the United States of America

# DEDICATION

This book is dedicated to my parents.

# Table of Contents

# ACKNOWLEDGMENTS

The front cover picture is printed with permission from Harris Corporation and **FLIR** Integrated Solutions. Thanks too for **DMT, LLC** (Detection Monitoring Technologies), **FLIR** Integrated Solutions, and **Vumii** for their permission to print other pictures within this book.

*Eddie R Hughes*

# Preface

As president and co-founder of a security radar manufacturing company, I have been asked many times over the years by prime contractors, integrators, and installers, "How do I perform site surveys for radar systems?" I have pulled together the relevant subject matter I have generated over the years to complete this site survey guide for the security specialist. Virtually every radar installation is accompanied by a high quality camera installation. Since cameras and radar are complimentary to one another in the security world, I have included site survey guidelines for cameras as well.

This book was written for the layperson. I have included only the technical knowledge that is required to be effective in conducting the site survey. Tables and figures often replace formulas. Numerous technical references on radar are available – some of which are listed in this book. I am also preparing a more comprehensive book on radar called, *Radar for the Security Professional.* This will be available sometime around the end of 2011.

This book provides site survey suggestions for all types of radar systems used in security as well as most daytime, thermal, and laser-illuminated cameras.   Mounting and tower considerations are discussed as well as grounding and communications.

Near the end of the book are example sites and my recommended solutions.   These examples include sites in cold, moderate and desert climates.

The final chapter in the book is concerned with the generation of the site survey report itself.   An outline and brief description are supplied.

There are a few conventions used in this book.   Any important definition, term, or phrase will be in bold and italics.   Whenever there is a rule of thumb on a page, the words rule of thumb will be present and will be underlined with a squiggle line.

All the rules of thumb that have been mentioned in this book are listed in Appendix A.   Rules of thumb can be dangerous when blindly applied to all applications.   But this book makes an effort to put conditions on how these rules are to be used.   Appendix B has a list of formulas for the more technical minded site surveyor.   And Appendix C has expanded information on some of the terms used in this book. Appendix D is some useful unit conversion information.

# 1 Introduction

## *Who Should Perform Radar and Camera Site Surveys?*

Whenever possible,  a site survey that includes radar and high-end cameras should be conducted by a security professional and qualified technicians from the sensor companies that are providing hardware. However, it may not be possible to have sensor engineers and technicians present.   Depending on the site location and budgeted cost, the site survey may need to be conducted solely by the security professional employed by the winning systems integrator or the customer.

The intent of this book is to give the reader a sufficient understanding of radar and camera technology and to supply the end user with a valuable and valid site survey.

People with engineering or science degrees will find the technical aspects of sensor site surveying to be an easy  study.   However, this type of background is not a requirement and this book makes an attempt to keep sensor science and mathematics easy to understand. The security professional should develop an understanding of:

- radar basics, including frequency, wavelength, spin rates, beam-width, environmental effects, false alarm abatement, mean-time-between-failure;
- thermal camera basics, including focal plane arrays, lenses, pan-and-tilt (PT) units and gimbals, wavelength, environmental effects, and cooled versus uncooled cameras;
- simple trigonometry (the very basics – most of which can be done using inexpensive calculators);
- power consumption and distribution;
- communication options;
- towers, poles and mounts;
- concept of Operations (CONOPS).

Getting the reader familiar with radar and camera technology is one of the two main goals of this guide. There are many other reference sources for understanding sensor technology and some are mentioned in this book. The second goal of this guide is to help the reader understand the essentials of optimum sensor placement.

Trigonometry formulas have been replaced with tables in this book, but an introduction to the most simplest formulae of sine, cosine, and tangent would seriously help any security professional that is not familiar with the terms. A cursory coverage of power, communications, and towers are also included in

*Figure 1-1. This is DMT's Dorado FMCW Marine Radar and the FLIR PT-606 camera. (Thermal camera courtesy of FLIR Systems, - learn more at www.flir.com/security.)*

this book. CONOPS are beyond the scope of this book, but there are many sources of information on the subject online.

## Certification

There are many organizations offering certifications for the security specialist. However, there are few courses that adequately train the lay person in site surveying for radar and cameras.

The publisher of this book offers radar and camera site surveyor classes for the security specialist. These 1-week classes feature a number of instructors (including the author) and focuses on both radar and cameras. These classes permit the site surveyor to see radar and thermal cameras in operation. Towers, command centers and communications are also covered in the instruction. A certificate of completion is provided to all participants.

Many times you can find great training courses available from the sensor manufacturers. Contact those manufacturers and ask when the next available class is being offered. Companies like the author's radar company, DMT, will allow site surveyors or systems integrators to sit in on classes if open spots are available. Certifications are provided at the end of these courses.

As great as these courses are, field experience cannot be substituted. Participation in site surveys in a variety of environments enhances the ability of the site surveyor. A widely traveled site surveyor that is familiar with the country or area to be surveyed is very valuable to any client. If the site surveyor is not familiar with the area, it is always good to talk to the customer or consult an expert in the area for insights into the potential threats to the site. Many radar and camera companies have points of contact that can assist in this matter.

## Things to Consider

Site surveying for sensors is an interesting and rewarding job. There are a few things to consider before you choose to do this for a full-time career.

Be respectful of potential hazards and know your limitations. If you have a fear of heights, then you will be less effective at many sites. Bring someone that can climb towers or go up ladders and take pictures for you. Many locations the author has surveyed had pre-existing towers or buildings that made effective sensor mounting points. Some of these towers and buildings, however, were no longer maintained for human habitation. Insects, snakes, rodents, birds and other animals sometimes live in these places. If you are afraid or disgusted by these things, then you'll need consider other career paths. Peeling lead paint, fiberglass, asbestos, and debris of all kinds can be present. If you don't come prepared, these can affect your long-term health.

A site surveyor must have or at least develop a good set of organizational skills. Site surveys are not cheap and documentation is crucial for a good report and recommendation. Coming to work with a good site surveying plan and then completing the reports as soon as possible on the return home (while the trip is still fresh in the mind) are essential for success.

New sensor technology and sensor enhancements are constantly being introduced in the security marketplace. The days of walls of camera feeds to sooth the watchman asleep are over. Faster radars, longer range cameras, lasers and LED illumination, and automation of security have been the latest developments by security sensor manufacturers over the last few years. One of the best ways to stay abreast of the changes in the market technology is to go to security shows. The big shows are essentials, such as ASIS (www.asisonline.org) and AUSA (www.ausa.org). If you can get in the FPED show in northern Virginia, which occurs every two years, it is highly recommended. The eighth FPED was in May 2011 and can be read about at www.fped8.com. Foreign shows, such as Milipol (http://en.milipol.com) in France and Qatar, SOFEX (www.sofexjordan.com), and the IDEX show (www.idexuae.ae) in the UAE are also recommended.

# 2 Radar Overview

## *Radar Introduction*

Radar stands for **RA**dio **D**etection **A**nd **R**anging.   All radars work by emitting radio waves.   Once transmitted, the radar begins to receive reflected energy from the environment and intruders.    The amount of energy received back at the radar's antenna can be truly minute.    For instance, a pulsed Doppler security radar with a transmitted peak power of 80 watts  will receive only 0.00000000000000000005 watts of power back from a person walking at 10 km (6.25 miles) away.

In order to understand radar, the site surveyor must understand some basic radar concepts and terms.   Appendix C has expanded definitions for radar.

Radar emits the energy with a certain repetition rate, bandwidth, and power level.   The combination of these are known as a **waveform**.   Radar systems can have more than one available waveform.    Since radar emits a radio wave, it will have a frequency.    **Frequency** is simply the number of cycles of a wave per second.   The span of transmitted frequencies of security radars is 1 to 24 GHz.   A **Gigahertz**, or **GHz**, is 1 billion wave cycles per second.   The majority of these security radars are at 9-10 GHz and  17-18 GHz for rotating radars. Non-rotating radars tend to radiate at 9 to 10,  and 24 GHz.  A radar's frequency is important in determining the radar's ability to see through weather and to detect small objects. This book does not cover see-thru-wall technology, which tends to be in the low Gigahertz frequencies.

Two things affect the size of the radar:

- maximum detection range,
- transmitted frequency.

So if the security radar is large (bigger than 5 feet in width), it is a good bet it is a long range radar and that it is probably transmitting at 9-10 GHz.

During World War II, radar development was classified and radar systems were designated using code names that represented a group, or band of frequencies. Each band was given a one to three letter designator. Most security radar systems are either X-Band (8-12 GHz), the upper side of Ku-Band (12-18 GHz) or the lower side of Ka-Band (24-40GHz).

Sometimes the term wavelength is used instead of frequency. **Wavelength** is related to frequency by the formula λ (wavelength in cm) = speed of light in cm/frequency in Hertz. A 10 GHz radar will have a wavelength of 3e+10/10e+9 = 3 cm.

Security radars come in many flavors. They can be grouped in long, medium and short range. They can also be divided into:

- 360 degree continuously rotating radar,
- Sector scan radars,
- Fixed beam radars (non-rotating).

360 degree rotating radar means the radar continues to spin in one direction. Security radars that are based on marine or shipboard radars spin only 360 degrees continuously. Some radars that spin 360 can also spin between any two angles, which is known as **Sector Scan**. In the last few years, a new set of sector scan radars that are non-rotating systems. These use **Phased Array Antennas** and can scan up to 120 degrees in azimuth. Phased array antennas electronically scan the beam. Although

this offers high *MTBF* (mean-time-between-failure) operation, the detection range is not uniform with azimuth. This is because the radar's beam widens as it nears the ends of the scan, which equates to lower gain, and therefore, shorter range. To counter beam spread, some vendors only offer 60 degree scan capability.

Fixed-beam radars radiate one or more beams of radar energy and monitor anything crossing those beams. These are sometimes referred to as "beam breakers." DMT has developed a line of these radars where the antennas are aligned in a semi-circle or circle for 180 to 360 degrees of coverage. Like phased array radars, these fixed-beam radars have no moving parts. No moving parts results in high MTBF.

There are five types of transmission technology used in the security radar market. These are:

- *CW*, or continuous wave (radar is always own)
- *FMCW*, or frequency modulated continuous wave (radar is always on, but the frequency changes over time)
- *Pulsed* (radar sends energy in bursts)
- *Pulsed Doppler* (radar sends energy in bursts, but also allows energy to be efficiently integrated and permits high precision speed measurement)
- *Pulse Compression* (radar sends energy in long bursts, and within the burst the frequency changes)

The differences between these radars can be summarized by looking at their waveforms. Figure 2-1 shows the *waveform* of each of these technologies. Pulsed and Pulsed Doppler waveforms "look" the same. (Doppler processing is the main difference between Pulsed and Pulsed Doppler radar.)

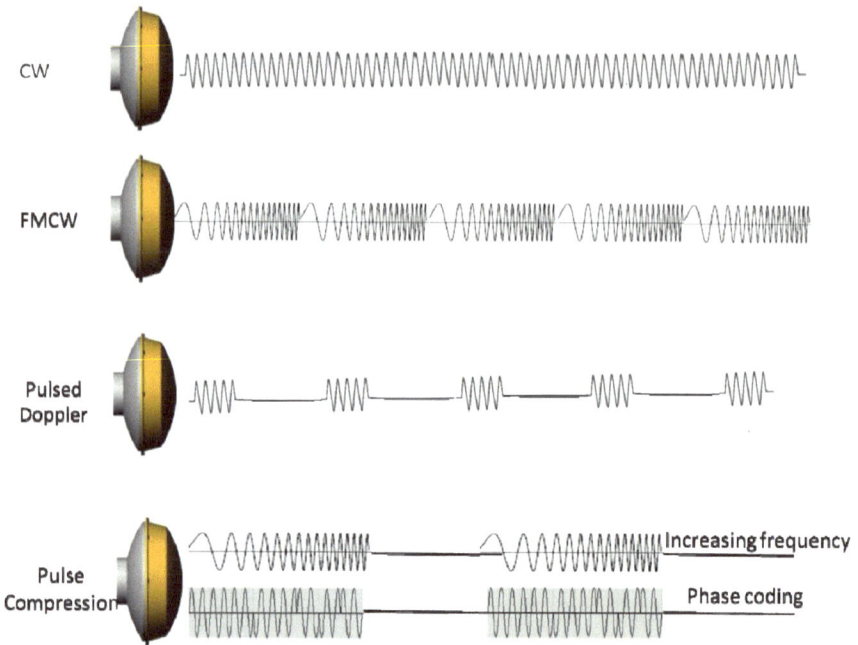

CW

FMCW

Pulsed
Doppler

Pulse
Compression                                      Increasing frequency

                                                  Phase coding

*Figure 2-1 . These are the most commonly radiated waveforms for today's security radar systems.*

Each of the types of radar mentioned have plusses and minuses. For instance, CW radars are the lowest cost radar and has the greatest range for the equivalent average power. However, CW radars cannot provide range to an object.

FMCW is one of the oldest forms of radar that has the range extent of CW. Unlike CW radar, it can provide range information. Although it is the simplest form of ranging radar, it can supply extremely high resolution and accuracy. It is the technology used in today's backup sensors on many cars, and is the most frequency used technology in security radar systems. But it can be easily jammed, is very sensitive to near-in reflections (which can render it blind at longer distances), and suffers from what is known as multipath, which are a form of false targets. Processing software can be used to overcome most of the multipath problems.

Pulsed radars perform best over water, but not as well over land. Pulsed Doppler radars are the most flexible radars and have the best performance over both land and sea of all single architecture radar systems. Unfortunately, they have slower update rates than FMCW due to the longer collection times required and the tremendous amount of processing they must perform. To compensate, pulse-pair or time domain processing can be used to speed up spin rates until there is a detection. Pulse Compression radars are like a mix of FMCW and Pulsed radars and are, therefore, better performing than either. But they are very complex, cost significantly more and have large minimum ranges.

Some companies, such as DMT, have started mixing technologies. DMT

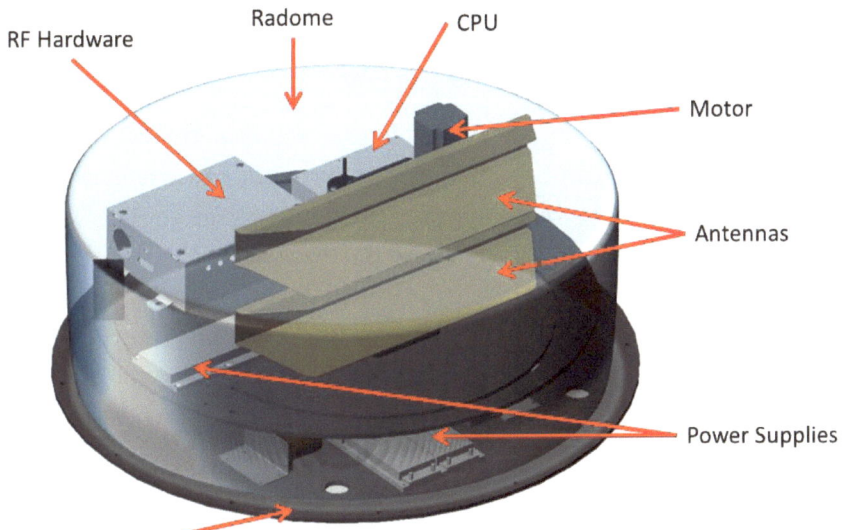

*Figure 2-2. This is an FMCW radar. This radar is typically built with 2 antennas -- one transmit and one receive.*

produces a radar called the HDR which has Pulsed, Pulsed Doppler and FMCW technologies all in one radar. It is extremely versatile and can meet any security requirement. It cannot be easily jammed and has extremely long mean-time-between-failures. But it is very expensive. Sometimes it is better to buy a radar the meets the mission requirements at a lower price and buy more of them than one radar that does everything.

## *Choosing the Right Radar*

If the security professional is asked to perform a site survey, it may be either a pre-solicitation survey or a post award survey.

Post award surveys are mainly a siting exercise conducted after the security contract has been awarded to an integrator, installer, prime contractor, or hardware vendor.   The security professional knows the type and quantity of hardware and must determine where to install it to provide the best   locations for towers and sensors to achieve optimal performance.

Pre-solicitation site surveys are surveys conducted for the customer to determine exactly what hardware is needed, how many, and an estimate their approximate location.   The security professional can give the customer a very good idea of the technology and requirements so that a request for proposal can be issued by the customer.  This type of survey requires an analysis of threats and their likely approaches to the site.

For the pre-solicitation surveys, the site surveyor needs to know which radar systems work best for a given site.  There are many variables to consider, including:

- Type of intruder,
- radar range,
- radar accuracy
- spin or update rate,
- operating temperature range,
- and all-weather performance.

Before you can decide which radar is best for your application, you must first determine the type of intruder you wish to detect with the radar.   Is it a human, a group of humans, vehicles, airplanes, boats, or combinations of these?   This is a very important question that is often not understood by those generating the requirements.   Remember that the smaller the intruder type (man versus vehicle, for instance) you design the solution around, the more expensive the solution becomes.

After intruder type, radar range is the next most important information to determine. Requirements can be generated solely on intruder type and the maximum detection range. The other four variables, radar accuracy, spin rate, operating temperature, and all-weather performance are secondary to intruder type and detection range. These two variables determine the majority of the cost of any radar selected.

It has become the norm to request that radars have very good accuracy in position and speed reporting of intruders. All security radars today have adequate accuracy. It is baffling to this author why so many countries and customers ask for 1 meter (or even less) accuracies. Most responders, such as military police, border guards and facility security guards would rather have reliable detection and tracking with no false alarms over knowing the intruders location to within a meter error. In discussions the author has had with security forces, even a 50 to 100 meters accuracy is sufficient given that the security response team must always search that amount of area anyway.

There also seems to be a misguided belief that positional accuracy will somehow allow the operator to count the number of intruders better. Resolution is what helps differentiate closely spaced objects. But if people are standing shoulder to shoulder, no radar on the market will be able to resolve them. This constant pressure to provide higher resolution and accuracy does not significantly increase overall performance, but drives the radar vendors to develop more expensive solutions to meet these accuracies without much real benefit.

Spin rates are important for understanding how fast the intruder's position will be updated. If the intruder is moving fast, but the radar is spinning slow, it may be too late to get a track established before the intruder has breached the perimeter.

Operating temperature is important, especially for very cold and very hot surroundings. The best performing radar system in terms of detection does no one any good if it is broken. Radars in the Middle East should have an operating temperature specification of at least +60 degrees C. Conversely, radars installed in cold climates like Norway, Canada, and Alaska should be able to survive at least -40 degrees C at startup.

All-weather performance implies the radar system works in all the conditions expected at the site. This is critical in understanding how many sensors are required at a site. If it is anticipated that a Ku-, K-, or Ka-Band radar will be used, then remember range can diminish quickly in fog, rain, snow or sandstorms. This means more systems may be needed to have good coverage in foul weather.

Although the number fluctuates, the more well-known security radar manufacturers are:

- DMT (Detection Monitoring Technologies)
- Detect Radar
- DRS
- EADS
- ELTA (a subsidiary of Israel Aerospace Industries)
- ICx (a FLIR company)
- Nav-Tech
- Plextek
- Raytheon
- Southwest Microwave
- Telephonics
- Thales

Each of these manufacturers produce one or more models of radar. Some specialize in certain technologies, such as ICx (FMCW, Ku-Band radar) and Southwest Microwave (fixed-beam, short-range Ka-Band radar). Others, such as DMT, EADS, ELTA, and Thales have broad selections of radar systems. The security professional should study all of the offerings of these companies so that intelligent decisions for a particular site can be made.

# *Intruder RCS*

*Radar Cross Section*, or *RCS*, is the measure of size of an object as perceived by the radar.  Its units are in either square meters (*sqm*, or $m^2$) or *dBsm* (dB square meters or decibel square meters).  dBsm is equal to 10\*LOG\*(RCS in square meters), where the LOG is the 10 Base LOG function.  In Excel, use the function LOG10() for this calculation.  dBsm Is a clever way to scale RCS so that small signatures and large signatures can be seen with equal clarity.   Unlike in optics, the RCS takes into consideration the material makeup of the object as well as its size.   Items that are made of metal will have a higher RCS than the identical item made of plastic.

A walking human RCS is normally taken to be 1 square meter, which is equal to 0 dBsm.   This is a  generalization or rule of thumb, that began many years ago and is not completely accurate.   In reality, the RCS of an average walking human will be more typically 0.4 to 2 square meters and will vary with viewing angle, vegetation, what the person is wearing or carrying, and movement.  "Average" humans are about 5'-10" in height and medium build.   The taller or larger the human, the greater their RCS will be.   And of course, the smaller or skinner the person, the lower their RCS will be.

Studies show that RCS is fairly consistent from 0.5 Ghz to about 10 GHz.  The higher the frequency, especially above 10 GHz, the greater the variability of human RCS will be.

When looking up RCS values for various intruder types, please remember that the any value listed as one number is usually the average value of RCS. Many are calculated as free space RCS. **Free space** means the object appears floating in a vacuum with no land or other surface nearby. So it's fairly unrealistic.

Empirical studies and in all radars tested by the author, a crawling human of any kind will have a lower RCS than a standing or walking human. Crawling humans come in two varieties: belly crawling and crawling on your hands and legs. The RCS for these will be less than 0.25 square meters. So this means a human crawling will be detected at a shorter range than a walking human in all environments.

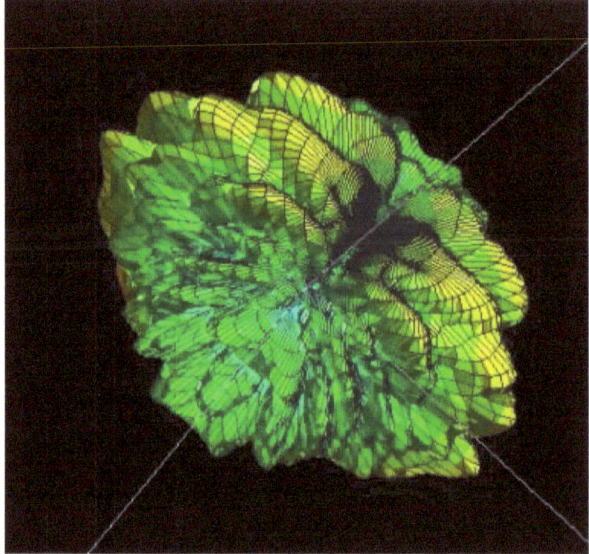

*Figure 2-3. The 3-D radar cross section of a standard, standing human is shown above.*

## *Radar Detection Range*

Radar detection range changes with the $4^{th}$ root of power or RCS. The means that if you know the detection range of a walking human from the radar vendor, then you can estimate the range for a crawling human by taking the $4^{th}$ root of the RCS ratio. For example, if a walking human is detected by Radar XYZ at 10 km, then a crawling human detection range will be:

(1) Range for walking human, or $R_{wh} = 10$ km.

(2) RCS for walking human = 1 sqm, and the maximum RCS for crawling human = 0.25 sqm

(3) Ratio of crawling human to walking human = 0.25/1 = 0.25.

(4) The 4$^{th}$ root is $(0.25)^{0.25}$ = 0.7071

(5) So for a crawling human, $R_{ch}$ = 0.7071 * $R_{wh}$ = 7.071 km. (This assumes the crawling human is in the line of sight of the radar.)

The detection range for a vehicle can also be calculated the same way. The RCS of a vehicle varies greatly from size, material used, shape, and viewing angle. A lightweight pickup truck will vary from 10 to over 100 square meters based on viewing angle alone. So if we take the worst case of 10 square meters, we can say the truck detection range will be:

(1) Range for walking human, or $R_{wh}$ = 10 km.

(2) RCS for walking human = 1 sqm, and the smallest RCS for a truck = 10 sqm

(3) Ratio of pickup truck to walking human = 10/1 = 10.

(4) The 4$^{th}$ root is $(10)^{0.25}$ = 1.778

(5) So for a truck, $R_{truck}$ = 1.778 * $R_{wh}$ = 17.78 km

Note that the range increased by a factor of 1.778 when the RCS increased by a factor of 10. This is a good rule of thumb to use for site surveys. The same rule of thumb can be used with transmitted power. It too is related to the 4$^{th}$ root of range. So if power were to increase by a factor of 10, the range would increase by a factor of 1.778. Suppose a radar vendor says that version 1 of Radar XYZ sees a walking person at 10 km and has 10 watts of power. If the same radar vendor offers version 2 of Radar XYZ with 100 watts of power, then the range for a walking human for Radar XYZ, version 2 = 10 km * 1.778 = 17.78 km.

As a side note, the above example of a 10 times power increase to get a 1.778 times improvement in range is why radar vendors often increase the size of their antennas instead. A 10 time power increase is costly; whereas, a larger antenna can be accomplished at a significantly lower cost. The larger the antenna, the more focused the energy and thus the longer the range.   This increase of energy from the larger antenna is a measure of its increased gain.   Range varies with the square root of gain. So an increase in the gain of an antenna by a factor of 10 (which is quite large) will result in an increase in the range by a factor of 3.16. Sometimes radar vendors will use both gain and power to increase range.

Be sure the selection of the intruder of interest makes sense before you begin the site survey.   For instance, what is the chance of a single human approaching a site in the middle of the desert?   The person would likely die from thirst or the environment before they could reach the protected site.    So in this case, you may wish to use a vehicle as the primary intruder.   A human will still be detected in this case, just not as far away. The same goes for a swimmer.   The likelihood of swimmer approaching a shore at five kilometers out is small.   It is more likely that a boat would drop the swimmer(s) or diver(s) off closer to shore. This is why jet skis or small boats are generally the specified intruder in requirements and in vendor specifications for seaside sites.

For security applications on land, a walking human is generally considered the threat of choice.   Water-based security applications will generally use a small boat or jet ski, since a swimming human is very hard to detect in choppy water.  A swimming human has less of an RCS than a crawling human, so even if the water is calm the swimming human will be seen at a small percentage of the distance of a boat.   The small boat will have a 1 to 15+ square meter RCS, depending on its size, shape and metal content. But remember that the amount of energy reflected by this intruder will vary greatly with sea conditions and orientation.   It should also be pointed out that the detection range of a boat is greater when moving than when at rest. This is because the RCS of the wake adds to the RCS of the boat and occupant.

Before performing a site survey for radar that will be scanning over land, obtain the reference RCS for the detection range listed for the radar in its specifications.  It will likely be a walking human.   If it is more than 1 square meter (0 dBsm), you should correct for this range to a 1 square meter target using the five-step method shown above.

For a radar that will be scanning over water, obtain the reference RCS or ask for the detection range for a small boat from the radar vendor. If the RCS number provided by the vendor for a small boat is greater than 10 square meters (10 dBsm), adjust it to 10 square meters using the five-step method just mentioned.

The application normally sets the range requirement. Typical applications for security radar systems include:

- Border security,
- Base security,
- Airport Security
- VIP (very important persons, which are typically heads of state and royal family) security,
- Oil and gas facility (including refineries) security,
- Pipeline security,
- Nuclear power facility or other similar alternative power facility security,
- Government test and operational sites (such as launch facilities and proving grounds) security,
- Waterside (river and seaside), port and harbor security,
- Railway security,
- And general wide area security.

The above applications can be broken down to their short, medium or long range applications. Table 2-1 lists the types of range performance expected for various security applications. For some applications, such as border security, more than one range requirement may exist.

| Table 2-1. Typical radar ranges used in various security applications. | | | |
|---|---|---|---|
| Security Application | Short Range (<=2 km range) | Medium Range (3-7 km range) | Long Range (>= 10 km range) |
| Border | | X | X |
| Base | X | X | |
| Airport | X | X | |
| VIP | X | | |
| Oil and Gas | | X | |
| Pipeline | | X | |
| Other Power | X | | |
| Government Sites | X | X | X |
| Waterside River | X | X | |
| Waterside Seaside | X | X | X |
| Railway Security | X | X | |

Unfortunately, many governments have not learned that the longer radar range requirement imposed on vendors may not necessarily result in long coverage ranges when installed. Without exception, the radar of choice for the majority of country border security **should** be radars with three to seven kilometer human detection range. For the small percentage of border that has longer ranges, the longer range radars should be considered. This will result in better security with proper overlapping coverage and greater probability of detection.

The practice of most governments, however, is to buy the longest range radar for the entire border. This results in the improper spacing in rough terrain, poor probability of detection, and overspending to say the least.

In planning for radars, the surveyor should understand the limiting factors of range. Most of the time, it is terrain or buildings that obscure line of site. But the surveyor should also understand that range can be limited by the curvature of the Earth. The point at which radar beam of energy is obstructed by the limb of the Earth is known as the **Radar Horizon** and can be easily calculated.

Figure 2-4 below gives tower height versus detection range and is a good tool for determining the maximum range a radar can detect. There is a good online calculator that can determine the radar horizon for any given tower height. At the time of this writing, the website address is **www.radarproblems.com/calculators/horizon.htm**.

Besides RCS, the other factors that determine radar range are:

- Average transmitted power,
- Antenna characteristics,
- Noise Figure of the radar,
- Dwell time,
- Environmental conditions.

We have mentioned already that transmitted power is related to the $4^{th}$ root of range. **Power** in this case refers to average power, not peak power.

# Radar Horizon

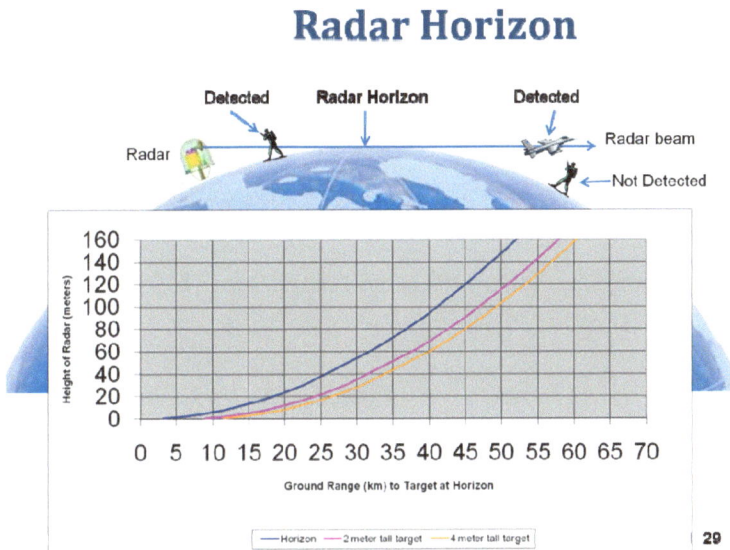

*Figure 2-4. The radar horizon is the distance where the Earth's limb blocks detection of objects beyond that distance.*

21

It doesn't really matter which radar technology we use for security, the same amount of transmitted average power is required to see an object at any given range. If the radar vendor provides only peak power, then contact them to obtain the average power. If it is a pulsed radar (Pulsed Doppler, Pulsed, Pulse Compression), the average power can be calculated using the formula:

**Avg Power (Watts)** = Peak Power (Watts) * Pulsewidth (sec) *PRF (Hz)

Pulsed radars continuously switch on and off their output signal. The length of time the signal is called the **Pulsewidth** and it is usually given in nanoseconds ($10^{-9}$ seconds) or microseconds ($10^{-6}$ seconds). The number of pulses per second is the **PRF (Pulse Repetition Frequency)** and it is usually given in units of Hertz (Hz) or Kilohertz (kHz). CW (Continuous Wave) and FMCW (Frequency Modulated CW) radars are always on, so their average power is equal to their peak power. There is a subset of FMCW radars known as **FMICW** (Frequency Modulated Interrupted CW). These radiate a sweeping frequency of a given bandwidth in bursts. It is similar to the transmitted waveform shown for pulse compression in Figure 2-1. The SIMRAD Broadband radar and the DMT Dorado marine radars are examples of FMICW radars. The average power radiated by FMICW systems is calculated:

**Avg Power**$_{FMICW}$ **(watts)** = Peak Power (watts) * sweeptime (seconds) * number of sweeps per second.

For more information about average power, please refer to Appendix C.

We have briefly mentioned already that antenna gain is related to the square root of radar range. So the bigger the antenna, the longer the detection range for any target will be. The size of the antenna determines the antenna's **Beamwidth**. The beamwidth (see Figure 2.5 and 2.6) is the width of the energy emitted from the antenna in units of degrees. The horizontal (also known as azimuth) beamwidth of the radar determines clutter power. ("Clutter" is discussed later in this chapter.)

Vendors should provide the horizontal and vertical beamwidths in their specifications. Request this if they do not.

Another important parameter is the antenna's **Sidelobes**. The sidelobes are the lower-power stray energy beams that are emitted by all radar antennas. Good performing radars will have antenna sidelobe energy that is less than 100 times less power (i.e., -20 dB) than the main beam of energy. The only effect of a sidelobe to radar performance is close-in to the radar if the antenna is designed well. If the sidelobe is too large, the radar will detect intruders in unintended locations (see Figure 2-7).

The antenna **Polarization** of the radar is also important. We could easily devote a whole chapter on polarization, but we will simply state here that it determines the orientation of the radio waves of the radar. Most radar systems are either vertically or horizontally polarized. Vertically polarized radars emit waves that oscillate up and down as it radiates through the air. Horizontally polarized radars emit waves that oscillate parallel to the ground. Vertically polarized radars will detect telephone poles and people better than horizontally polarized radar. Horizontally polarized radar will detect long swells on the ocean, ships, and cars better than vertically polarized radar.

The top drawing in Figure 2-5 is a cartoon depicting the antenna beamwidth as is often drawn in site survey diagrams. The width of the beamwidth is measured at the half-power (3dB down) points. The bottom figure is a theoretical pattern that might be the design goal for a radar. It is very common to see this in a text book and it was created using an antenna modeling software package.

A real antenna pattern that was measured is shown in Figure 2-6. The real measured pattern is what the site surveyor should request from the manufacturer. The circular plot is known as a polar plot, which shows

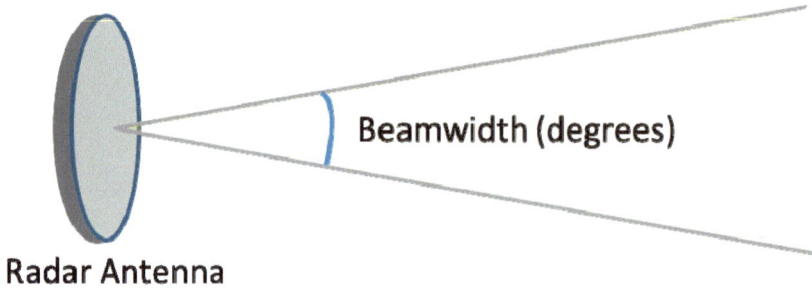

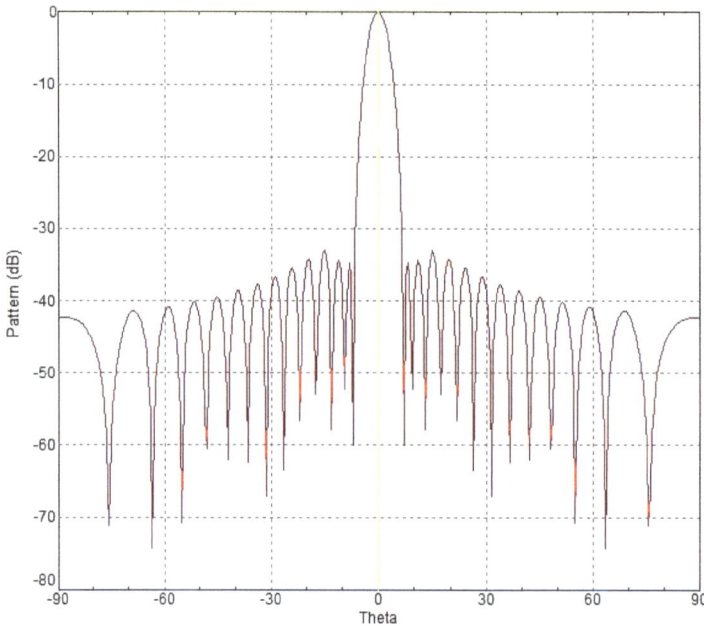

*Figure 2-5  The radar antenna pattern (radiated shape) is typically shown as a pie-shaped main beam, as shown in the top figure. This represents the width of the half-power (3dB down) beamwidth.   The bottom graph is more indicative of a true radar pattern.  Theta is the angle relative to the boresight of the antenna.  The main beam (main lobe) is the central lobe at zero degrees theta and each of the outlying lobes are the sidelobes.  Note the sidelobes are more than 35 dB down from the main lobe. If lobes are higher on the graph, say -10 to -20 dB, there is a chance for erroneous detection of objects at angles other than the boresight.*

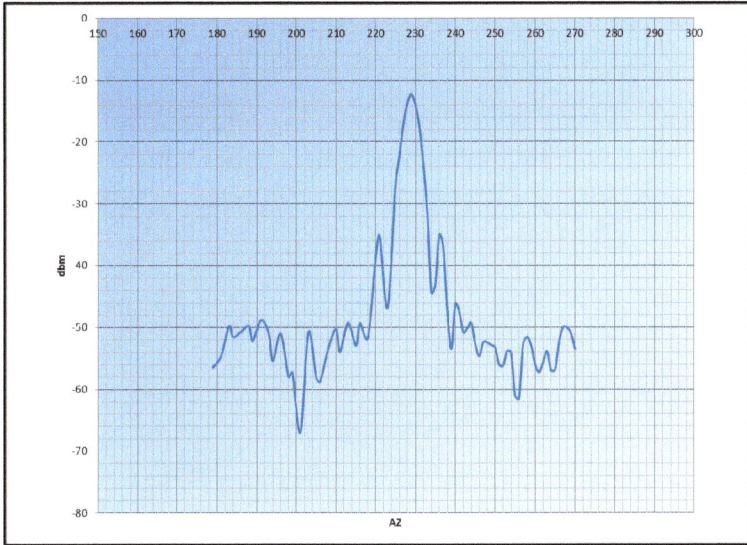

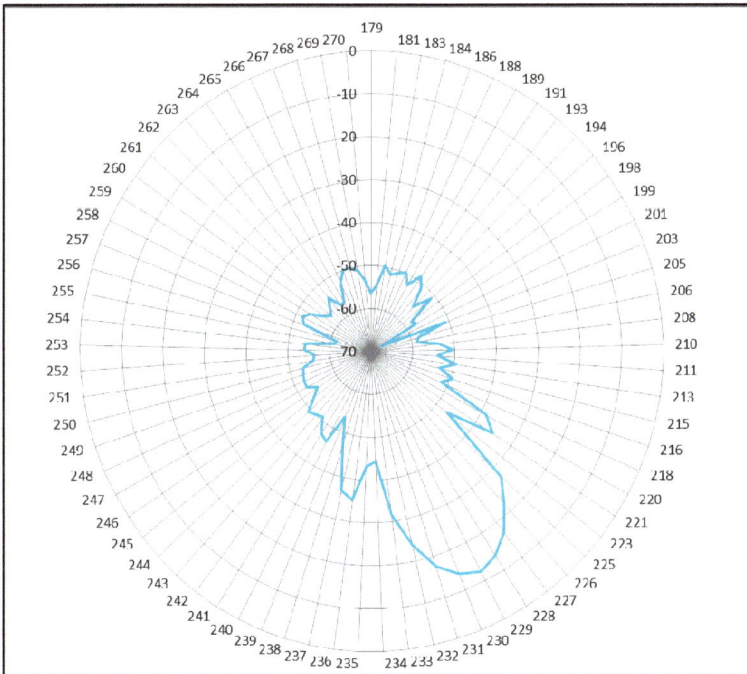

*Figure 2-6. The top and bottom graphs are measured azimuth antenna patterns. The top pattern is in Cartesian coordinates and the bottom is a polar plot. On the polar plot, each ring represents a 10 dB step.*

how the pattern actually looks when radiated. Just the top part of the ***main lobe*** (at the 3 dB down, or half-power points) of the antenna is normally listed in specifications. But there is significant strength radiated beyond this half-power points.

Main Beam Power = 80 W
Sidelobe Power = 0.4 W
RCS Bus = 1000+ square meters

Result:
- Bus detected in sidelobe
- Angular position of bus reported at 0 degrees instead of 8 degrees

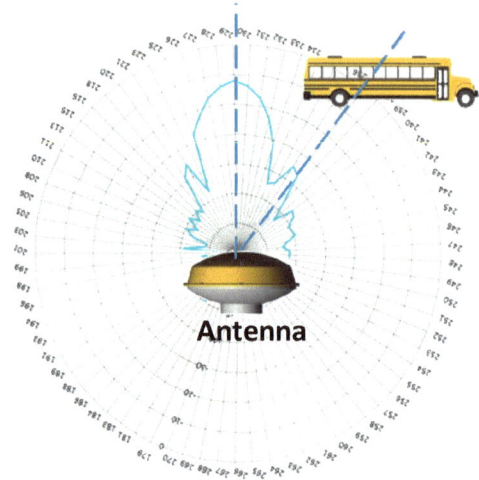

Antenna

*Figure 2-7. This figure shows that sidelobes are very important in intruder detection. Sidelobes that are too high would result in inaccurate reporting of range. Even with -23 down sidelobes in the above antenna pattern (which is good), the bus signature is large enough to be detected in the sidelobe. However, since motor positions are taken at the main lobe angular position, the error in the case above would be 8 degrees in angular position.*

Figure 2-8 shows a radar antenna mapped on a cut-away drawing of a tower. This 3-pole tower is similar to a Rohn 45G (or larger) tower. You can print this on a transparency and the place it over a map of the area or facility in which the survey is being conducted. It illuminating just how much energy radiates into unwanted areas of a site.

Figure 2-9 are examples of different antennas used in security radars. The large antenna has the most gain and longest range. The smallest antenna (in the center) has the least gain and shortest range.

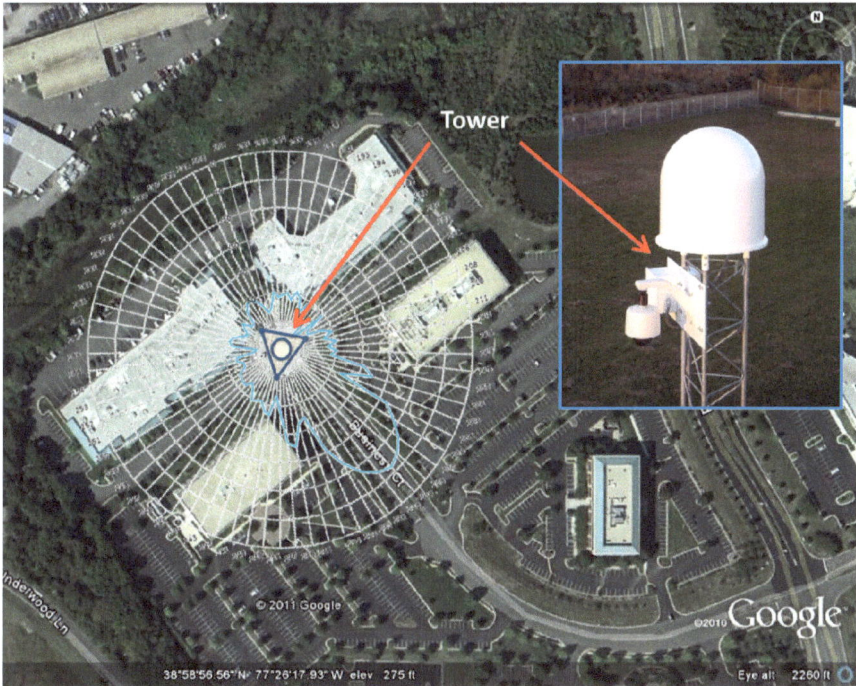

*Figure 2-8  This is a Google Earth map showing an overlay of a radar antenna pattern and a cut-away drawing of a Rohn tower.*

The **Noise Figure** is a measure of the strength of the internal noise of the radar.  The greater the noise figure, the greater the reflected signal from objects must be to detect them.   A noise figure of most well designed radars should be less than or equal to 5 dB (decibels).

**Dwell Time** is the amount of time the radar accumulates return energy from the object of interest.   In pulsed and Doppler radars, this is a critical parameter.   The longer the dwell, the greater the radar's detection range will be.     Unfortunately, the longer the dwell the slower the radar will spin.    If the spin rate drops too slow, the tracker performance (which affects positional and speed reporting accuracy) will suffer.

**Environmental Conditions** affect the radar range of EVERY radar.   Some radars are impacted much more by

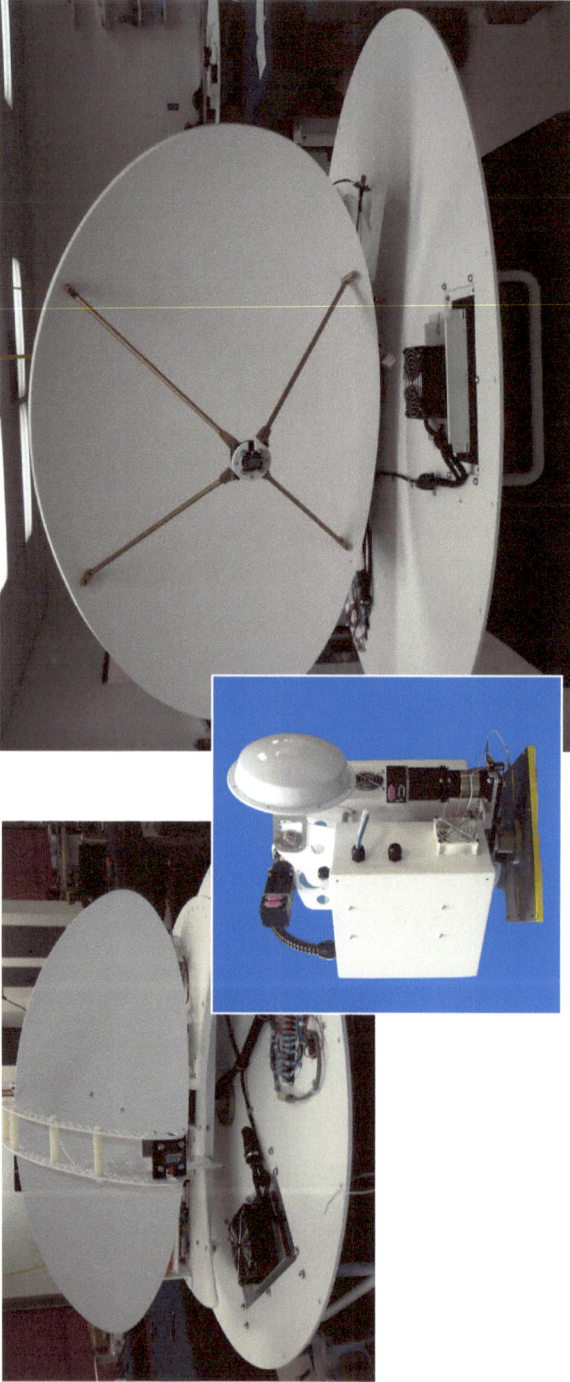

*Figure 2-9. The right picture shows a large antenna for long range detection. Since the antenna is elliptical, the pattern emitted is elliptical. It has a narrow horizontal beamwidth and slightly wider elevation beamwidth. The left picture shows an antenna that is a little smaller, so both the horizontal and vertical beams are bigger. The left antenna has a larger ratio between vertical and horizontal dimensions and has what is called a "fan" beam due to its similarity to an Asian hand fan. The inset center picture shows a radar with a smaller antenna and is therefore, a short range radar (1.5 km for humans). The antenna is circular so its pattern similar to a flashlight beam.*

the environment than others.    Radars that are impacted less by the weather are:

- L-Band, S-Band, C-Band and X-Band radars (radars less than 10 GHz)
- Doppler radars

Land radars that are less impacted by vegetation blowing are:

- L-Band, S-Band, C-Band and X-Band radars (radars less than 10 GHz)
- Doppler radars

Radars looking over water that are less impacted by swells, choppy water, currents, etc. are:

- L-Band, S-Band, C-Band and X-Band radars (radars less than 10 GHz)
- Doppler radars

It is clear there is a theme here.    The good rule of thumb is that lower frequency radars -- from about 1 GHz to 10 GHz -- are much less impacted by any type of environmental condition.    Actually frequencies below 1 GHz are also great in all environmental conditions, but security systems are not built at these frequencies due to the large antennas that would be needed to adequately direct the radar beam.

There are a number of security radars that reside in the 17-18 GHz region. Environmental conditions, such as rain, snow, sandstorms, and even fog will have a noticeable impact on these radar systems performance.    This is not necessarily a bad thing for all applications.    For instance, if the facility is a VIP location or a military base,  the grounds are normally flat and well cleared.    As a result, the higher frequency systems operate just fine and the site surveyor's only concern is to make sure there is sufficient margin in range to account for the losses due to weather.

It is important when preparing for a site survey to understand what sources of false alarms and nuisance alarms will be present. False Alarm is a term that is misused and sometimes abused in the security world when it comes to radar. A radar *False Alarm* is an alarm generated from one or more of the following:

- internal noise in the radar and/or external noise from solar or other radiation or emitting sources;
- improper installation, such as being too close to a wall, or the tower itself;
- improper setup (too large of a pulsewidth or a spin rate that is too fast);
- unknown source.

In the case of the tower above, a tower that is not rigid can generate false alarms by its sway or rotation or both in high winds.

Nuisance Alarms are often mistakenly called false alarms. When talking to a radar engineer, do not get these terms confused. *Nuisance Alarms* are alarms caused by real objects in the coverage zone, such as:

- trees and telephone poles
- blowing vegetation
- fans, windmills, wind turbines
- gates and doors
- items blowing in the wind, such as fences, windsocks, etc..

Radars vendors remove nuisance alarms by using clutter maps or clutter notches (if they are Doppler radars) or both. *Clutter* is the detectable return energy from the environment that is not the intruder or object of interest. Grass, trees, and other foliage are the most common sources of radar clutter.

A *Clutter Map* is a three-dimensional (x(meters), y(meters), z(signal strength in dB)) representation of the return energy from the coverage area of a radar when there are no real intruders present. Think of it as a reference that should be updated often by the radar. All new detection updates of the coverage area are preceded by the subtraction of the clutter map from the new return. This leaves only significant changes, which could be the presence of a new intruder.    The remaining signals are passed to a detection algorithm and then a tracker.

Doppler radar systems can employ a clutter map, but they also have more information available that can be exploited to  remove clutter.  Clutter notches are one item that can be used.   A *Clutter Notch* is the removal of a signal frequency or group of frequencies from the Doppler Spectra in one or more look directions.

Doppler radars work by measuring the rate of change of the transmitted frequency caused by moving objects.  Doppler radar can measure from 0 Hertz (cycles per second) up to about a few kilohertz of shift.    This amount of frequency shift is directly related to the speed of the objects in the radar's beam.    The pulse-repetition-frequency (PRF) sets the maximum amount of frequency shift and radial speed of a radar.    For instance, a X-Band radar with a 4 kHz PRF rate can detect speeds of 0 to about +/-72 mph.    (A minus sign indicates the object is moving away from the radar, the + sign means the object is moving toward the radar.) Some radar vendors that are unfamiliar with Doppler radar claim that they cannot see stationary objects.  **This is untrue; it is a myth.**  A Doppler radar can measure stationary objects because the very first shift measured is 0 Hz (= stationary objects).

The beauty of Doppler radars are that they can remove clutter (such as grass blowing in the wind) by simply ignoring (notching) any detections at these slow speeds (see Figure 2-10).    Doppler radar algorithms work in frequency space, so energy that is periodic (such as ocean swell) can be removed without removing intruder signal.    Doppler radar is covered in much more detail in the *Radar for the Security Professional* book mentioned earlier.

Avoiding false and nuisance alarms are the primary concern of the site surveyor. The proper tower configuration helps reduce sources of unwanted alarms.

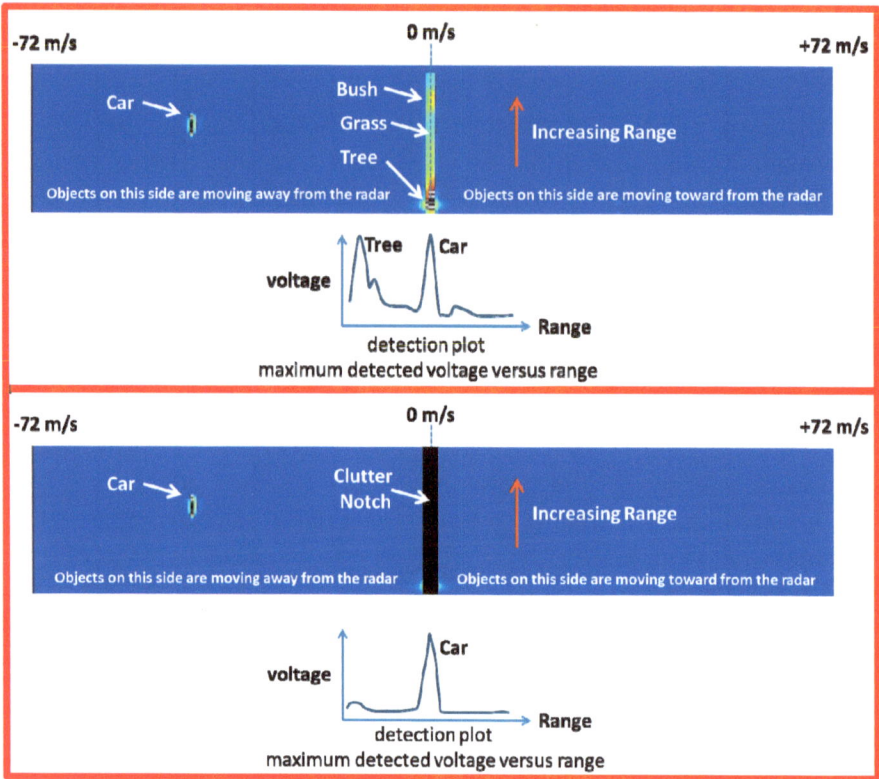

*Figure 2-10. Doppler plots and Line plots. The blue Doppler plot at the top shows the signatures of a car and various clutter sources, such as a tree. The line plot below it shows the maximum voltages detected for increasing range. The bottom Doppler plot shows the signature of the car, but the clutter notch masks the signatures of other clutter sources. The bottom line plot shows the results of the clutter notch – only intruders of interest are detected.*

# References

These references will provide a complete understanding of radar and most the necessary formulas to determine performance of any vendor's product.

Books:

- ***Radar Principles, Technology, Applications***, by Byron Edde Prentice Hall
- ***Radar Principles for the Non-Specialist***, by J.C. Toomay and Paul J. Hannen, SciTech Publishing
- ***Radar for the Security Professional***, by Eddie Hughes, Deep Sea Publishing
- ***Radar Reflectivity of Land and Sea,*** by Marice W. Long, Artech House, Inc.
- ***Radar Cross Section,*** by Eugene F. Knott, John F. Shaeffer, Michael T. Tuley, Artech House, Inc.
- ***Introduction to Radar Systems***, by Merrill I. Skolnik, McGraw-Hill Book Co.
- ***Radar Technology Encyclopedia,*** by David K. Barton, Sergey A. Leonov, Artech House, Inc.
- ***Radar System Analysis and Modeling***, by David K. Barton, Artech House, Inc*.*

Links:

- *http://www.radartutorial.eu/index.en.html* (a great site for radar education)
- *http://www.vectorsite.net/ttradar_1.html* (this is another great tutorial site that covers the basic principles of radar and historical perspectives)
- *http://en.wikipedia.org/wiki/Radar* (Wikipedia has a great site here for radar, antennas, and a good radar band table)
- *http://www.answers.com/topic/radar* (includes definition and a host of articles from encyclopedias and other open source literature.)
- *http://www.explainthatstuff.com/radar.html* (encyclopedia-like information)

# 3 Camera overview

There are far more security camera manufacturers than radar manufacturers. A partial list of the more well-known security camera companies include:

| | |
|---|---|
| Argon ST (a Boeing Company) | L-3 (Cincinnati Electronics) |
| Aventura | Obzerv |
| Axis | Opgal |
| Axsys (a General Dynamics Company) | Panasonic |
| BAE | Patronus Labs |
| Bosch | Pelco |
| Cohu | PVP Advanced EO Systems |
| FLIR | SIM Security & Elec. Sys |
| GyroCam (a Lockheed Company) | Sony |
| ICx (a FLIR company) | Vicon |
| IEC | Vumii (an Opgal Company) |
| Klymar | Wescam (an L-3 company) |

There are likely many missing camera company manufacturers in the above list, so search security shows for vendors not listed above. These camera manufacturers make a wide variety of camera products. This book does not cover the site survey basics for low-cost security cameras (domes or fixed mounted cameras) – there are just too many applications where these may be employed and the number of vendors are in the hundreds. The list of camera manufacturers provided above build rotating cameras with at least a kilometer range in daylight or 500 meters at night. Similar to radar manufacturer's, some of the companies on the above list build very specific camera technology.

The primary camera technologies used in security are:

- CCTV (closed-circuit TV) for daylight or low-light applications,
- Thermal Cameras for nighttime, daytime and light fog and smoke applications, which include:
    o Cooled Passive IR (infrared),
    o Uncooled Passive IR,
    o Uncooled Active IR,
- Active illuminators, including laser-illuminated and LED-illuminated camera technology for daytime and nighttime applications.

Camera zoom levels are often used to give the impression of high quality. However, the ***field of view*** is the best measure of performance. The field of view is very similar to beamwidth for a radar. It provides the angular area over which the visual image will extend for the camera. The smaller the field of view is, the larger an object will appear at a given distance. Very long range cameras will have field of views like 0.55 x 0.41 (width x height) degrees. Short range cameras will have field of views like 17x13. Chapter 4 gives a useful table for field of view. Camera ranges should be selected to be less than or equal to the range of the radar used for the site. Having a camera with longer range is a waste of money, because it will require manual pointing to see objects at these ranges.

The spin rates for any camera used for security is an important parameter when used with a radar. After years of experience with camera control, the author's radar company found that a 30 degrees per second spin rate is a minimum performance requirement. Cameras that can spin this rate or higher will be able to make it to most intruder detection positions reported by the radar. Any delays associated with communications will push this number higher.

# CCTV

**CCTV** cameras provide high quality video images during daylight hours. Many modern CCTV cameras are actually **CCD** (charge-couple device) cameras. These have a two-dimensional array of elements that are sensitive to light. Today's CCD camera range from 1/16" to 4/3" in size, with 1/4", 1/3", and 1/2" being the most common in security cameras. Some models, like the Sony Super HAD camera used in many of the security dome cameras, is also low-light sensitive. It can see very well with only minimum illumination or even near-infrared.

A CCTV camera's ability to see in low light will be directly related to its *lux* rating. Lux, or illuminance, is the number of lumens per square meter. It is the measure of intensity of illumination on a surface. A footcandle is another measure of illuminance used in the USA. 1 footcandle is equal to about 10 lux. Now most quality color cameras have lux ratings of about 0.3 to 1 for 1/60 second exposure. Sub-$3,000 low-lux security cameras can have ratings as low as 0.00018 lux for 1/2 second exposures.

CCTV cameras come in hundreds of models, so it is beyond the scope of this book to cover all these models. But there is one thing to be very aware of when purchasing these – operating temperature specifications. Most have very limited operating temperature ranges. Most CCTV cameras are well under +50 degrees Celsius operating temperature range, which is inadequate for the southwest of USA or the Middle East unless mounted in the shade or cooled in some way.

# Thermal Cameras

Thermal cameras are a good solution for nighttime viewing. The cameras are designed to see in the infrared spectrum. Heat, such as body heat or vehicle motors, are great sources of infrared emissions. Today's thermal cameras use focal plane arrays (FPA's) that consist of a plane of pixel detectors. The difference between a FPA and CCD is the wavelength of light to which they respond. Good thermal cameras have thermal arrays consisting of a minimum 320 (horizontal ) x 240 (vertical) pixels. Expensive longer range thermal cameras will have 640 x 480 or 640 x 512 FPA's.

Viewing range on most thermal cameras are based on the ***Johnson Criteria*** or something similar. The Johnson Criteria has three range categories: Detection, Recognition, and Identification. "Detection" means that there is a minimum of two pixels illuminated across 0.75 meters in the object plane. "Recognition" means that there are six pixels illuminated and "Identification" means that there are twelve pixels illuminated. The important point here is that Identification is not used in the classic sense of the word. Identification, or ID here means sufficient pixels are illuminated to allow the operator to tell this is a soldier with a gun. There is insufficient information to say this is Jim or Amy. So when analyzing camera performance and certainly when planning for camera siting, a good rule of thumb is to use the Recognition range. The Detection ranges quoted will disappoint your client if you use these distances for siting at critical locations. See Appendix C for more information on the Johnson Criteria.

Thermal cameras come in single, dual, three, four or continuous field of views. Price increases accordingly. A single field of view thermal camera cannot zoom, unless the zoom level is digital. Digital zoom is not very useful. Dual field of views will have two zoom levels, which are often labeled Narrow and Wide. Three or triple field of view cameras have three zoom levels often labeled Narrow, Mid and Wide. Continuous field of views started to arrive in force in the security market on very high-end thermal cameras in 2009. When the user zooms with a continuous field of view thermal camera, it looks like the zoom of a CCTV (daylight) camera.

Passive (non-radiating) thermal cameras come in cooled and uncooled versions. Most cooled cameras are currently LWIR (Long-Wave IR) and most uncooled are MWIR (Mid-Wave IR). LWIR use 3-5 micron detectors while MWIR are 8-12 micron detectors. Cooled cameras are long range cameras, but are sometimes undesirable because of the short times to failure. The failures are almost always due to cooler failure. Coolers will often fail at a rate of two times per year for desert climates. Beginning in 2010, the author saw some cooled thermal camera's MTBF specifications exceeding one year. Uncooled cameras top out at about 3 km detection range. They are more reliable and have a MTBF (mean-time-between-failure) of well over 18 months.

Focal length is an important factor that affects range. The higher the focal length for the camera, the longer the range. 25 mm will be a relatively short range, while 1000 mm will be a very long range. For cooled thermal cameras, it is common to see focal lengths of 250 up to 2000 mm. For uncooled cameras, the focal lengths are typically between 25 to 135 mm. There are a few high-quality uncooled cameras with up to 300 mm focal lengths.

Most thermal cameras that are now offered with a daylight camera as well. The minimum performance of the camera is always the nighttime performance. Daytime performance will usually be significantly better. When a daylight camera is supplied, they may have lux rating up to 2.0 lux since the thermal camera will switch on when it is night.

Active thermal cameras use IR emitters are normally used in very short range applications for the most part. Over the last year, there have been some camera integrators that have added this capability to some very long range passive cameras to extend range. In researching for this book, the author found very few of these cameras being built by the large companies and most did not have wide operating temperature ranges.

There are a number of good sites for thermal camera information:

- *http://www.thermalvideo.com/thermal-imaging-systems/flir_vsr-6.htm* has a good set of pictures shows the difference between images that are at Detection, Recognition, and Identification ranges.
- *http://en.wikipedia.org/wiki/Thermal_camera* has a good discussion on cooled and uncooled thermal cameras.
- *http://www.infraredtraining.com/* is a good site to visit if you wish to be trained on thermal technology and be certified.


## *Laser-illuminated Cameras*

Laser-illuminated camera entered the market a few years ago and have expanded quickly. At the time of this writing, laser-illuminated camera manufacturers include:

- Kylmar (a General Dynamics company),
- Obzerv,
- Patronus Labs,
- PVP,
- SIM-Danis,
- Vumii (purchased by Opgal).

These camera systems use low-light CCD cameras and that use a laser to illuminate the surroundings. They do not follow the Johnson Criteria, but they do use the Detection, Recognition, and Identification terminology. However, in the case of laser-illuminated cameras, the terminology is more accurate. Detection for laser-illuminated cameras is more like Recognition for thermal cameras. Recognition for laser-illuminated cameras is far better than Identification in thermal cameras.

Unlike thermal cameras, the laser-illuminated cameras cannot see through fog and smoke. So their use for maritime applications should be limited to shorter ranges unless coupled with a thermal camera.

One of the better features of laser-illuminated cameras is that they achieve ranges well beyond those offered by uncooled thermal cameras. Since they do not use coolers, they have greater MTBF (mean-time-between-failure) specifications than cooled thermal cameras. All long range laser- and LED-illuminated cameras are continuous zoom systems. The majority of thermal cameras are still one, two or three field-of-views (i.e., 1 to 3 zoom levels). Those thermal cameras that are continuous zoom are substantially more expensive. Also, in the USA, most laser-illuminated cameras fall under Commerce jurisdiction. Whereas, most cooled thermal cameras fall under ITAR (US State Department).

LED-illuminated cameras are similar to laser-illuminated cameras with the obvious exception – they use Light Emitting Diodes (LED's) instead of a laser. LED illumination has been used for many years to give fixed (non-rotating) cameras the ability to see at night. But in most cases, they were limited to 100 feet. In the last couple of years, there has been a marked increase in longer range LED illuminators. There are accomplished used higher intensity LED arrays. At the time of writing, the maximum ranges for LED illumination with a rotating camera is about 1000 meters.

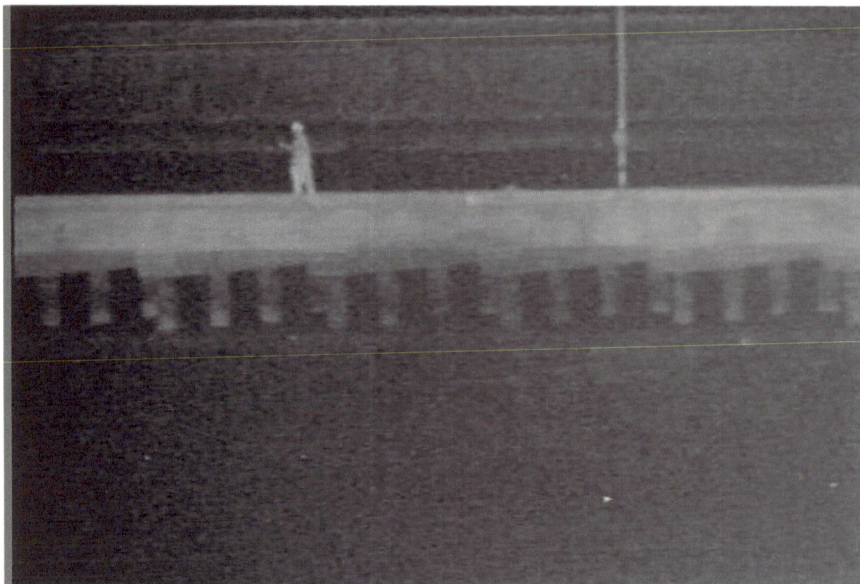

*Figure 3-1. A human walking on a moonless night as seen by a laser - illuminated camera.*

Unlike thermal cameras, which are limited by pixel density, target type, and detector dynamic range, laser-illuminated and LED-illuminated cameras are limited by the range of the laser more than anything else. And the laser strength and even the type of laser is influenced by the need to be eye-safe. All laser-illuminated cameras used for security are eye-safe when used as instructed by the manufacturer, but they do have restrictions.

The wavelength of the laser, the lens, and its output power affect the stand-off safety range of the laser-illuminated camera. The most common wavelength used is around 800 nm. The power output for a 3000 meter range laser-illuminated power will typically be 10 or more watts. The stand-off distance for safe operation is 100 meters for a 3000 meter range laser-illuminated camera. For a 1000 meter range camera, the stand-off safe range is about 50 meters. The safe range for any laser-illuminated camera will be provided by the manufacturer and is normally found in their marketing literature and manuals.

The stand-off safety range is measured from the laser and not the ground distance. Tower mounting the cameras will effectively shorten the ground range to the point of safe operation.     A requirement of the installation will be to set the minimum negative (downward) tilt angle of the camera.  This will prevent people close to the tower from accidental eye damage.     Use the formula below to calculate the safe ground distance from the camera tower.

$$C = Square\ Root(A^2 - B^2\ )$$
*where, A = manufacturer's stated safe range in meters*
*B = tower height in meters*
*C = safe ground distance from tower in meters*

You can also calculate the lowest allowed tilt angle of the camera for the range you calculated above using the formula:

*Lowest Tilt Angle (in degrees) = ArcTan (B /C)*

LED-Illuminated cameras are safe to the eye at all ranges and do not have the same restrictions as the laser-illuminated.

The last point to make on laser- and LED-illuminated cameras are their daytime ranges.  It is not uncommon that the daytime performance may be several thousand meters.  For instance, the Vumii Claritii (a LED-illuminated camera) has a 500 meter night-time performance and a 5000 meter daytime performance against a human.

## *Camera Conclusions*

No one camera works for all installations and missions.  The selection of the camera is always site-specific.    The best solution is to mix technologies whenever possible.   All-weather performance is best provided with thermal cameras.     Identification and through-glass performance is best provided by laser- and LED-illuminated cameras. Having daylight, thermal and illumination technology on a site is a great way to ensure the best possible detection of intruders in most situations and sites.

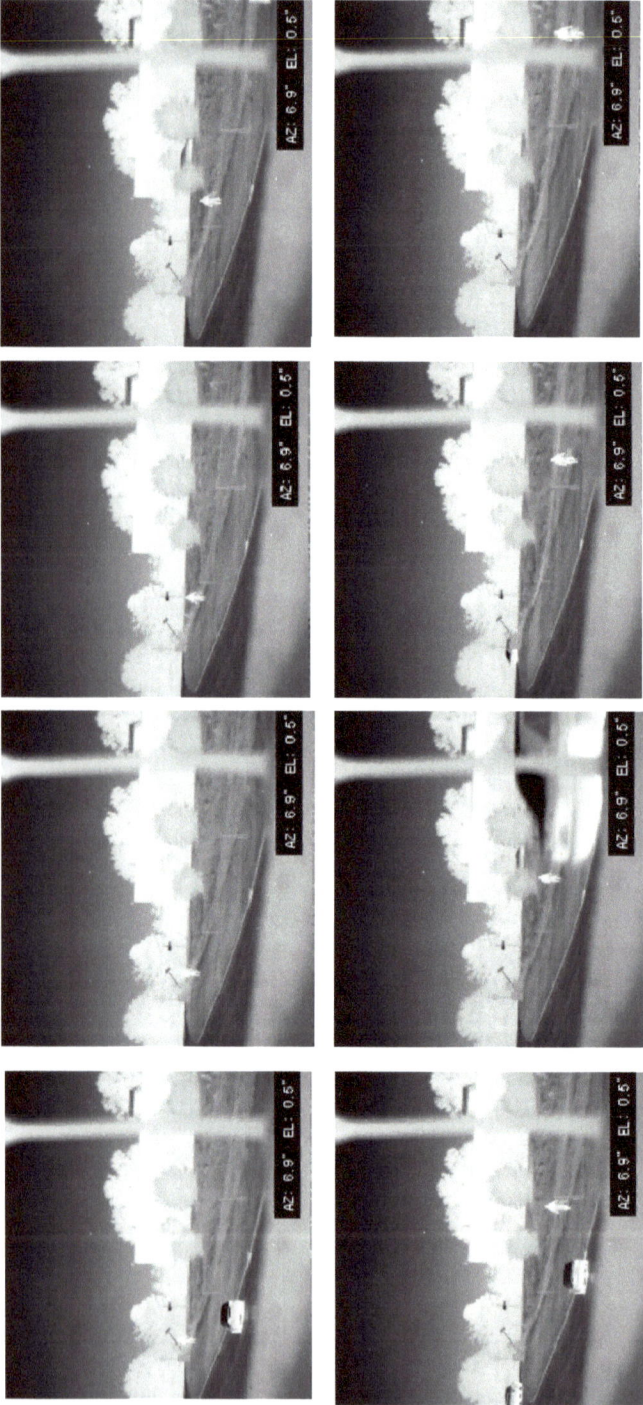

*Figure 3-2. This is a sequence of thermal images of a walking man from the recognition range to identification range. These are white hot images (white = heat) of the author walking with shorts in the summer in Virginia.*

# 4 Before the Survey Begins

## *Tools of the surveyor*

It is assumed that the reader is not employed by a radar and camera company, but has been assigned the responsibility to conduct a site survey for these types of sensors. Site surveying is best served with an actual site visit. Seeing the site and any obstacles that may cause obstruction or restrict sensor line of sight is always preferred over an electronic survey using satellite images, maps and pictures.

The general site survey can be conducted with as little as one person. But experience has shown that two people work better at spotting issues.

Items needed for a site survey:

- binoculars,
- calculator,
- camera or camera phone,
- handheld GPS or GPS enabled smart phone,
- compass,
- tape measure and laser range finder (or measurement wheel if not along the water),

- printed maps of the facility from programs such as Google Earth or Windows Live,
- this book.

The binoculars are useful for sites where long-range radars are going to be installed, for looking at mounting points on existing tall structures, and for riverside or hard to reach locations. Small 10 power (10x) binoculars are perfectly fine and can fit in your pocket. The calculator helps in the calculation of tilt angles and in beamwidth estimates in meters. And the tape measure (minimum 25 foot length) helps to determine distance of a potential site to nearby obstructions. A laser range finder also reduces time and walking distances. It is a necessity for riverside locations. Inexpensive laser range finders will see to about 1000 feet, while slightly more expensive ones are available with ranges of more than 1000 meters. If a laser range finder is not available, then a measurement wheel is the next best substitute for getting fairly accurate measurements of large spaces.

Do not start any site survey without first studying the site from a detailed map and a satellite image. Google Earth and Windows Live Maps are very good tools for this. Print a copy of the map to bring along on the survey, and make sure the map scale is visible on the printed copy.

Table's 4-1 and 4-2 are useful tables in conjunction with the map. Table 4-1 gives the surveyor the radar antenna coverage in meters for various ranges in kilometers. For example, suppose the surveyor must plan for an invisible fence of radar energy around a VIP residence (as in Figure 4-1). Let us assume, for example, that the surveyor has determined that two multi-beam radars should be pointed at each other along the 500 meter wall near the street to the north in Figure 4-1. (A multi-beam radar is one that has multiple antennas that can be pointed in different directions. This can also be accomplished using a multiple single-beam radar systems). The distance between the wall to the street that runs parallel to the wall is 60 meters, so you have decided a 50 meter wide swath along the wall is adequate. What beamwidth will work for this application? Looking on the chart under 0.5 km for the wall length, which is the third column, the closest beamwidth that results in a 50 meter wide coverage is 6 degrees. Table 4-2 provides the width in meters of the field-of-view (FOV) at various ranges for cameras.    Note that although a 6 degree FOV is adequate for the camera, it is wise to be a little wider if the camera has no pan (spinning) capability.

A camera or a camera phone is extremely important for site surveys. Sometimes they are not allowed, but insist as much as possible.   Keep notes on each picture taken.   Get a latitude and longitude for each reading.   Use the compass for taking a reading relative to True North for each picture to make sure of your vantage point and viewing angle. These pictures will be very useful in preparing a great site survey report. Be sure to correct for magnetic variance (also called magnetic variation), which will affect the compass readings.   For example, the magnetic variance correction for Washington, DC is about 10 degrees.   For much of the Middle East, the correction for magnetic variance is near 0 and can therefore be ignored.   Wikipedia has a good description of magnetic variance (*http://en.wikipedia.org/wiki/Magnetic_declination*).

A good estimation of magnetic variance can be obtained at the NOAA website: *http://www.ngdc.noaa.gov/geomagmodels/Declination.jsp* The calculator requires a latitude, longitude and date.

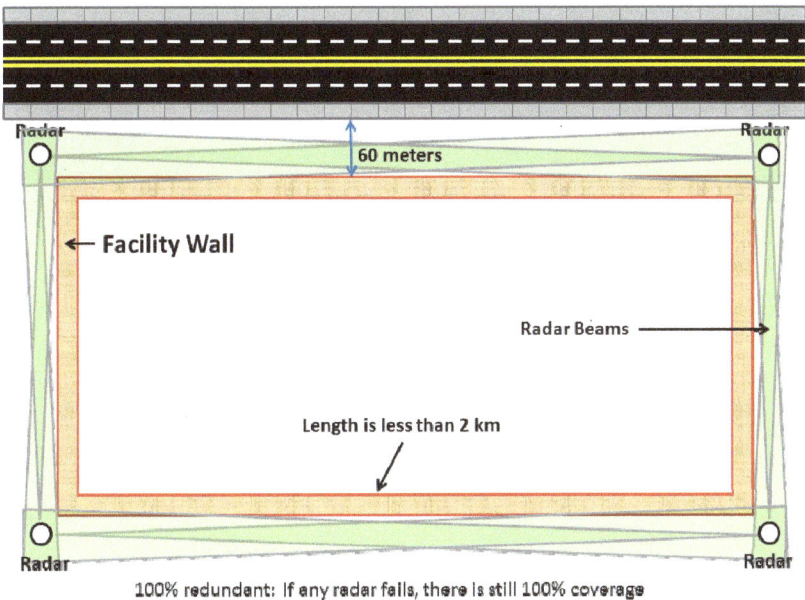

*Figure 4-1. Example security scheme for a VIP residence using 4 multi-beam radar systems. Each radar is emitting two beams along the walls of this residence.*

Table 4-1: Width of beam in meters for various ranges in kilometers (km)

| Beamwidth (deg) | 0.3 | 0.5 | 0.8 | 1 | 1.5 | 2 | 3 | 5 | 7 | 10 | 12 | 15 |
|---|---|---|---|---|---|---|---|---|---|---|---|---|
| 1 | 5 | 9 | 14 | 17 | 26 | 35 | 52 | 87 | 122 | 175 | 209 | 262 |
| 2 | 10 | 17 | 28 | 35 | 52 | 70 | 105 | 175 | 244 | 349 | 419 | 524 |
| 3 | 16 | 26 | 42 | 52 | 79 | 105 | 157 | 262 | 366 | 524 | 628 | 785 |
| 4 | 21 | 35 | 56 | 70 | 105 | 140 | 209 | 349 | 489 | 698 | 838 | 1047 |
| 5 | 26 | 44 | 70 | 87 | 131 | 174 | 262 | 436 | 611 | 872 | 1047 | 1309 |
| 6 | 31 | 52 | 84 | 105 | 157 | 209 | 314 | 523 | 733 | 1047 | 1256 | 1570 |
| 7 | 37 | 61 | 98 | 122 | 183 | 244 | 366 | 610 | 855 | 1221 | 1465 | 1831 |
| 8 | 42 | 70 | 112 | 140 | 209 | 279 | 419 | 698 | 977 | 1395 | 1674 | 2093 |
| 9 | 47 | 78 | 126 | 157 | 235 | 314 | 471 | 785 | 1098 | 1569 | 1883 | 2354 |
| 10 | 52 | 87 | 139 | 174 | 261 | 349 | 523 | 872 | 1220 | 1743 | 2092 | 2615 |
| 11 | 58 | 96 | 153 | 192 | 288 | 383 | 575 | 958 | 1342 | 1917 | 2300 | 2875 |
| 12 | 63 | 105 | 167 | 209 | 314 | 418 | 627 | 1045 | 1463 | 2091 | 2509 | 3136 |
| 13 | 68 | 113 | 181 | 226 | 340 | 453 | 679 | 1132 | 1585 | 2264 | 2717 | 3396 |
| 14 | 73 | 122 | 195 | 244 | 366 | 487 | 731 | 1219 | 1706 | 2437 | 2925 | 3656 |
| 15 | 78 | 131 | 209 | 261 | 392 | 522 | 783 | 1305 | 1827 | 2611 | 3133 | 3916 |
| 16 | 84 | 139 | 223 | 278 | 418 | 557 | 835 | 1392 | 1948 | 2783 | 3340 | 4175 |
| 17 | 89 | 148 | 236 | 296 | 443 | 591 | 887 | 1478 | 2069 | 2956 | 3547 | 4434 |
| 18 | 94 | 156 | 250 | 313 | 469 | 626 | 939 | 1564 | 2190 | 3129 | 3754 | 4693 |
| 19 | 99 | 165 | 264 | 330 | 495 | 660 | 990 | 1650 | 2311 | 3301 | 3961 | 4951 |
| 20 | 104 | 174 | 278 | 347 | 521 | 695 | 1042 | 1736 | 2431 | 3473 | 4168 | 5209 |
| 25 | 130 | 216 | 346 | 433 | 649 | 866 | 1299 | 2164 | 3030 | 4329 | 5195 | 6493 |
| 30 | 155 | 259 | 414 | 518 | 776 | 1035 | 1553 | 2588 | 3623 | 5176 | 6212 | 7765 |
| 35 | 180 | 301 | 481 | 601 | 902 | 1203 | 1804 | 3007 | 4210 | 6014 | 7217 | 9021 |
| 40 | 205 | 342 | 547 | 684 | 1026 | 1368 | 2052 | 3420 | 4788 | 6840 | 8208 | 10261 |
| 45 | 230 | 383 | 612 | 765 | 1148 | 1531 | 2296 | 3827 | 5358 | 7654 | 9184 | 11481 |
| 50 | 254 | 423 | 676 | 845 | 1268 | 1690 | 2536 | 4226 | 5917 | 8452 | 10143 | 12679 |
| 60 | 300 | 500 | 800 | 1000 | 1500 | 2000 | 3000 | 5000 | 7000 | 10000 | 12000 | 15000 |
| 70 | 344 | 574 | 918 | 1147 | 1721 | 2294 | 3441 | 5736 | 8030 | 11472 | 13766 | 17207 |
| 80 | 386 | 643 | 1028 | 1286 | 1928 | 2571 | 3857 | 6428 | 8999 | 12856 | 15427 | 19284 |
| 90 | 424 | 707 | 1131 | 1414 | 2121 | 2828 | 4243 | 7071 | 9899 | 14142 | 16971 | 21213 |
| 100 | 460 | 766 | 1226 | 1532 | 2298 | 3064 | 4596 | 7660 | 10725 | 15321 | 18385 | 22981 |
| 110 | 491 | 819 | 1311 | 1638 | 2457 | 3277 | 4915 | 8192 | 11468 | 16383 | 19660 | 24575 |
| 120 | 520 | 866 | 1386 | 1732 | 2598 | 3464 | 5196 | 8660 | 12124 | 17321 | 20785 | 25981 |

Table 4-2: Width of the field-of-view in meters for various ranges in kilometers (km)

| FOV (deg) | 0.3 | 0.5 | 0.8 | 1 | 1.5 | 2 | 3 | 5 | 7 | 10 | 12 | 15 |
|---|---|---|---|---|---|---|---|---|---|---|---|---|
| 0.4 | 2 | 3 | 6 | 7 | 10 | 14 | 21 | 35 | 49 | 70 | 84 | 105 |
| 0.6 | 3 | 5 | 8 | 10 | 16 | 21 | 31 | 52 | 73 | 105 | 126 | 157 |
| 0.8 | 4 | 7 | 11 | 14 | 21 | 28 | 42 | 70 | 98 | 140 | 168 | 209 |
| 1 | 5 | 9 | 14 | 17 | 26 | 35 | 52 | 87 | 122 | 175 | 209 | 262 |
| 1.2 | 6 | 10 | 17 | 21 | 31 | 42 | 63 | 105 | 147 | 209 | 251 | 314 |
| 1.4 | 7 | 12 | 20 | 24 | 37 | 49 | 73 | 122 | 171 | 244 | 293 | 367 |
| 1.6 | 8 | 14 | 22 | 28 | 42 | 56 | 84 | 140 | 195 | 279 | 335 | 419 |
| 1.8 | 9 | 16 | 25 | 31 | 47 | 63 | 94 | 157 | 220 | 314 | 377 | 471 |
| 2 | 10 | 17 | 28 | 35 | 52 | 70 | 105 | 175 | 244 | 349 | 419 | 524 |
| 2.2 | 12 | 19 | 31 | 38 | 58 | 77 | 115 | 192 | 269 | 384 | 461 | 576 |
| 2.4 | 13 | 21 | 34 | 42 | 63 | 84 | 126 | 209 | 293 | 419 | 503 | 628 |
| 2.6 | 14 | 23 | 36 | 45 | 68 | 91 | 136 | 227 | 318 | 454 | 544 | 681 |
| 2.8 | 15 | 24 | 39 | 49 | 73 | 98 | 147 | 244 | 342 | 489 | 586 | 733 |
| 3 | 16 | 26 | 42 | 52 | 79 | 105 | 157 | 262 | 366 | 524 | 628 | 785 |
| 4 | 21 | 35 | 56 | 70 | 105 | 140 | 209 | 349 | 489 | 698 | 838 | 1047 |
| 5 | 26 | 44 | 70 | 87 | 131 | 174 | 262 | 436 | 611 | 872 | 1047 | 1309 |
| 6 | 31 | 52 | 84 | 105 | 157 | 209 | 314 | 523 | 733 | 1047 | 1256 | 1570 |
| 7 | 37 | 61 | 98 | 122 | 183 | 244 | 366 | 610 | 855 | 1221 | 1465 | 1831 |
| 8 | 42 | 70 | 112 | 140 | 209 | 279 | 419 | 698 | 977 | 1395 | 1674 | 2093 |
| 9 | 47 | 78 | 126 | 157 | 235 | 314 | 471 | 785 | 1098 | 1569 | 1883 | 2354 |
| 10 | 52 | 87 | 139 | 174 | 261 | 349 | 523 | 872 | 1220 | 1743 | 2092 | 2615 |
| 12 | 63 | 105 | 167 | 209 | 314 | 418 | 627 | 1045 | 1463 | 2091 | 2509 | 3136 |
| 14 | 73 | 122 | 195 | 244 | 366 | 487 | 731 | 1219 | 1706 | 2437 | 2925 | 3656 |
| 16 | 84 | 139 | 223 | 278 | 418 | 557 | 835 | 1392 | 1948 | 2783 | 3340 | 4175 |
| 18 | 94 | 156 | 250 | 313 | 469 | 626 | 939 | 1564 | 2190 | 3129 | 3754 | 4693 |
| 20 | 104 | 174 | 278 | 347 | 521 | 695 | 1042 | 1736 | 2431 | 3473 | 4168 | 5209 |
| 24 | 125 | 208 | 333 | 416 | 624 | 832 | 1247 | 2079 | 2911 | 4158 | 4990 | 6237 |
| 28 | 145 | 242 | 387 | 484 | 726 | 968 | 1452 | 2419 | 3387 | 4838 | 5806 | 7258 |
| 32 | 165 | 276 | 441 | 551 | 827 | 1103 | 1654 | 2756 | 3859 | 5513 | 6615 | 8269 |
| 34 | 175 | 292 | 468 | 585 | 877 | 1169 | 1754 | 2924 | 4093 | 5847 | 7017 | 8771 |
| 36 | 185 | 309 | 494 | 618 | 927 | 1236 | 1854 | 3090 | 4326 | 6180 | 7416 | 9271 |
| 38 | 195 | 326 | 521 | 651 | 977 | 1302 | 1953 | 3256 | 4558 | 6511 | 7814 | 9767 |
| 40 | 205 | 342 | 547 | 684 | 1026 | 1368 | 2052 | 3420 | 4788 | 6840 | 8208 | 10261 |

The elevation profile for each sensor site will be needed. There are three ways the site surveyor can determine the amount of elevation change that is required for all azimuth scan angles:

- use a surveyors transit to determine angles;
- use Google Earth or similar programs to measure altitude points and then calculate the elevation angles;
- use a handheld GPS and walk or drive the outer perimeter.

Surveyor's transits are very easy to use and are accurate. You can use it for both azimuth and elevation angle measurements. Simply level it and the rest comes easy using your eye to align the scope objects in the distance. The author searched online and found many perfectly good, used transits for under $50.

Google Earth is an excellent tool for measuring elevation angles. Once you zoom into the map of the survey area, use the ruler function to give range between the potential sensor site and the longest range covered by the radar and/or camera. As the cursor is dragged over the map, the bottom of the screen shows the updated latitude, longitude and altitude. Choose to save the line and then right mouse click the line and select "show elevation profile." A pop-up graph will appear below the map with the elevation profile (see Figure 4-2). Be sure that you zoom in on the map in the vicinity of the line you have drawn. The elevation profile will have better resolution when the map is at maximum zoom. Once the profile has been drawn, the map can be zoomed back out as desired. As the mouse is moved over the elevation profile, a corresponding arrow will indicate the map position of that elevation.

Use the windows calculator or any handheld calculator and compute the elevation angle using the formula:

***Angle in degrees = ArcTan (altitude difference in feet/range in feet)***

ArcTan (or Inv Tan or Tan$^{-1}$) on the Window's calculator is accomplished by pressing "Inv" and then "tan" buttons. Make sure the radial button "Degrees" is selected.

Most handheld GPS units are capable of saving waypoints. Some will even save elevation profiles over time. Either function can be used to obtaining elevation angles using the same formula mentioned above. The author's handheld GPS will give an elevation or altitude profile as one moves. This is very useful as you walk the perimeter of any site.

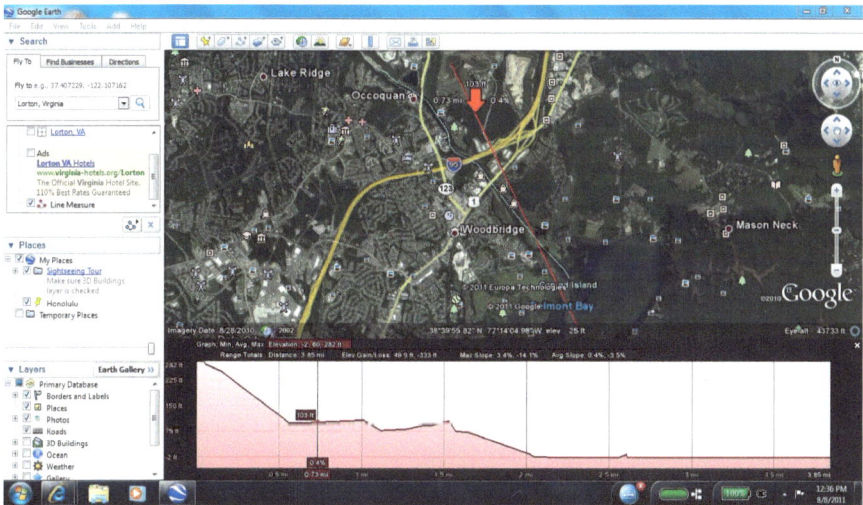

*Figure 5-2. This is a screen capture from Google Earth. The red line represents the look direction of the radar. Using the ruler tool of Google Earth, the site surveyor can determine the elevation profile of the look direction. Moving the cursor on the elevation profile causes the red arrow to move alone the red line.*

## Mobile Phones with Built-in GPS

Today's mobile phones offer many benefits to the site surveyor. Most smart phones come with a built-in GPS. It can sometimes take a little effort to find the actual latitude and longitude coordinates, but most come with this feature in the standard software that comes with the phone or offer applications (apps) that can be downloaded with the capability.

In addition to the GPS, most also come with cameras.  And if you have bought a smart phone from 2010 and on, your photos are being tagged with geolocation information.  This is great for the surveyor because his location information is automatically recorded with his pictures.

The image format used by smart phones and many other digital electronics is called the exchangeable image file format, or ***exif***.  To see if your phone has this capability, look at an outdoor photo you may have taken in the photo gallery or the album.  Select the photo and then go to Menu and either Details or Properties.

If it doesn't show the coordinates, load the digital photo onto your PC and check to be sure you actually have the feature.  On a Windows PC, right click on the photo and select Properties.   Then select the Details tab.  Scroll down until you get to the GPS section where you will then find the coordinates if the phone stores them.   If you own a MAC, then open the photo in Preview.  Then click Tools and select the Inspector tab.   Go to the GPS section where the coordinates for the photo are stored (if enabled in the phone).

Your phone may have this feature disabled.   You will obviously need to enable it to be able to use this feature.

If you use this smart phone capability in lieu of bringing a separate GPS, you should take at least one picture at the exact location that you anticipate the radar and camera to be.   It is a good idea to take a 360 degree set of pictures from each sensor location.  You can then stitch these pictures together to give you the full view at each tower's location.  This has been extremely helpful to the author on his site surveys.  Panoramic movies captured at each tower site are also useful.   Especially if you give a presentation afterwards.

## *What to wear?*

When going on a site survey, comfort and safety are essential.   Given the number of items mentioned at the beginning of this chapter, a backpack is a good storage idea for items if the site is large.  A water bottle, a raincoat, sunglasses, work gloves (for climbing ladders, rusty rails, and dirty doors or openings) and a hat should also be included in the backpack as is necessary. Cargo pants are useful for carrying some of the items that you use often during the site survey. If your company has a logo shirt and it is appropriate, wear this so that you can be easily identified.

If the site is an oil refinery, gas facility, or industrial site, the site surveyor will often be required to have:

- a reflective vest,
- steel-toed boots,
- safety glasses (high contrast or tinted are good),
- hard hat.

Although the items above (except for steel-toed boots) may be available by the security department of the site being visited, having your own will fit substantially better and is more professional looking. If steel toed boots are not required, comfortable shoes (not sneakers) are recommended. No flat-sole shoes should be worn to these sites. There should be good treads on any shoe worn on site surveys.

If there are old building, towers, or other structures are being surveyed for potential sensor mounting points, be sure to bring a dusk mask. Don't skimp on the quality of these masks. Be sure to buy a form-fitting mask with at least a nose bridge clip. Dusk respirators are much better for extended use situations. These are tight-fitting masks with an exhalation valve. By allowing hot air that is exhaled to leave the mask, the mask becomes more comfortable and stays much dryer.

## Identification and Documentation

Having the right identification and personal documentation will make the site survey go faster and reduce problems encountered at the sites. Not having the right ID can prevent the site survey from happening.

Research the site before visiting it.   Find the all the necessary points of contact (POC) at the site that are needed to gain access.  Plug the POC's mobile and office telephone numbers into your phone before departing to the site.   Once signing in at the security office, the security officer may require you to contact the POC for escort.   This preparation can reduce the amount of time lost in the sign-in process.

There are two forms of identification that help greatly reduce problems in gaining entry into an industrial facility in the USA or a USA base overseas:

- TWIC -- Transportation Worker Identification Credential,
- Military ID, CAC (Common Access Code, a DoD access card).

The TWIC card can get you into almost any petrochemical site in the USA. It is also recognized in many other sites considered to be essential for national security.   To apply for a TWIC card online, go to the website: https://twicprogram.tsa.dhs.gov/TWICWebApp/PrivacyAction.shtml.
The CAC card is only available to US military personnel and US military contractors.

The other forms of identification that the site surveyor should bring are:

- Driver's license,
- Passport and any Visa if overseas,
- Medical history and list of medicines,
- Contract and customer contact information,
- Your company ID,
- Business Cards.

The medical history and medicine list is important in the event you are injured while performing the site survey.

Before starting the site surveyor, plan a brief meeting to the onsite security officer and staff, or site manager.   Bring a paper brief and an electronic version in the event they have an LCD projector. The security staff or site manager can be helpful in recommending site locations, in understanding power and communication availability, and operations.

## *Risk Assessment*

One of the main purposes of a site survey is a ***risk assessment***. Risk assessment is basically the analysis of all threats to the site. The quality of the assessment can greatly affect the project costs.

The site surveyor should evaluate the site's surroundings before the physical inspection begins. Determine local traffic patterns by foot and vehicle and choose the ones that pass closest to the site for your physical inspection. The author and his staff at DMT have actually parked for hours watching and documenting patterns for several days at sites. This included weekdays, weekends, days, and nights. This is not always necessary, but can be very illuminating for busy sites.

# 5 Radar Site Surveys

There are 7 items that need to be covered during a site survey:

(1) Confirmation of line of site over secured area;
(2) No main beam blockage;
(3) Confirmation of elevation coverage;
(4) Overlapping azimuth coverage;
(5) Determination of security perimeter or boundaries;
(6) Ensure the radar horizon limitation is met;
(7) Consider the radar accuracy .

Each site survey should begin by locating sensor sites that supply adequate sensor coverage.   All modern radars are line-of-sight (LOS) sensors, which means they can only detect objects visible by cameras or the human eye.   Use binoculars to verify there are no obstructions.   Tall bushes, trees, poles and small buildings all cast "shadows" behind them. Radar *shadows* are areas directly behind large poles, trees, rocks, hills, buildings or other objects where the radar coverage is absent or extremely weak.   The effect is similar to fast moving water in a stream flowing around a rock.    Adjust the radar site to avoid these objects or recommend their removal.   The radar energy normally wraps around trees and poles up to about 3 inches in diameter for X-Band and about 1-2 inches for Ku-Band radars.[1]  However, large diameter poles, buildings and large, wide brushes can generate shadows all the way to the outside perimeter boundary.   Deep trenches or gullies are also areas that should be examined carefully during site surveys.

Care must also be taken if the radar must scan over close-by obstructions that are lower than the radar.   For instance, radar towers installed near buildings should be substantially taller than the building.   The building will not only cast a radar shadow, it may cause large reflections of the antenna sidelobe energy.   If the building is close to the radar, the reflections of the antenna sidelobes off the building could saturate the radar receiver and cause false alarms or reduce range or both.   If the building is further away, the radar shadow will be large.

A radar installed near along a wall can have its range severely limited if part of the beam is touching the wall   Range reduction occurs due to the absorption or redirection of the radar energy by the wall.   If the energy is reflected away by the wall, then nuisance alarms from significant distances off the wall may be reported as an intruder at the wall.    To make sure this does not happen, follow this procedure:

- Determine the distance x, where x = the maximum distance the radar must detect along the wall.
- Using the following rule of thumb formula, determine the wall stand-off distance for the radar:

*Minimum distance of radar from a wall = x * tan(Θ/2)*

- Table 5-1 supplies the above formula for various ranges. Table 4-1 and 5-1 are related by 1/2 the beamwidth.   (Note: Table 4-1 is the width in meters of the beamwidth at various ranges and is a measure of the cross-range error of the radar.)

---

[1] The pole must be in the far-field of the radar – see Appendix B.  If the pole is close (i.e., in the near-field), the pole should be less than one wavelength thick.

- The above calculated range ensures that the radar will have reduced false and nuisance alarms and maximum range when installed near a wall. Please note that this is for radars mounted inside or outside the wall perimeter. Also remember that some liberty can be taken with the above calculation if there is plenty of open area directly in front of the wall.

Windblown small bushes and trees will generate nuisance alarms for many systems – especially for non-Doppler radar systems. Windmills, large trees and tree lines, large power poles and power lines can also produce high-magnitude nuisance alarms. Make sure to point out any vegetation that should be removed from the site. If the client does not want to remove a potential false alarm generator, then reposition the radar to see around these objects.

Some radar software can blank out (ignore) areas that generate unwanted alarms. So if a bush is causing a nuisance alarm, the radar operator or administrator can set the radar display to ignore these alarms. Radars, such as those sold by DMT, have display software that can define the radar's coverage area graphically. This is accomplished by drawing the perimeter on a map of the facility using the mouse. So unwanted alarms from tree lines, for instance, can be avoided by drawing the protection perimeter just inside the tree line by a few meters.

The surveyor must also determine the elevation profile for each potential radar site. For most sites the radar is run with elevation = 0 degrees, which means perfectly level. There is no antenna tilt up or down as it is spinning. But if there are elevation changes that must be made, then the intended radar's antenna tilt range should be considered. When the terrain varies as the radar spins in azimuth, the elevation or tilt angle of the antenna may need to be continually adjusted. A radar that adjusts its antenna tilt angle as it spins to account for terrain conditions is known as a ***terrain following*** radar.

| Beamwidth (Degrees) | Table 5-1. Closest Distance Between Sensor and Wall for VIP, Prison, and Fenced or Walled Sites in Meters | | | | | | | | | | | | |
|---|---|---|---|---|---|---|---|---|---|---|---|---|---|
| | Length of Wall in Meters | | | | | | | | | | | | |
| | 50 | 100 | 150 | 200 | 250 | 300 | 350 | 400 | 450 | 500 | 600 | 800 | 1000 |
| 1 | 0.4 | 0.9 | 1.3 | 1.7 | 2.2 | 2.6 | 3.1 | 3.5 | 3.9 | 4.4 | 5.2 | 7.0 | 8.7 |
| 2 | 0.9 | 1.7 | 2.6 | 3.5 | 4.4 | 5.2 | 6.1 | 7.0 | 7.9 | 8.7 | 10.5 | 14.0 | 17.5 |
| 3 | 1.3 | 2.6 | 3.9 | 5.2 | 6.5 | 7.9 | 9.2 | 10.5 | 11.8 | 13.1 | 15.7 | 20.9 | 26.2 |
| 4 | 1.7 | 3.5 | 5.2 | 7.0 | 8.7 | 10.5 | 12.2 | 14.0 | 15.7 | 17.5 | 21.0 | 27.9 | 34.9 |
| 5 | 2.2 | 4.4 | 6.5 | 8.7 | 10.9 | 13.1 | 15.3 | 17.5 | 19.6 | 21.8 | 26.2 | 34.9 | 43.7 |
| 6 | 2.6 | 5.2 | 7.9 | 10.5 | 13.1 | 15.7 | 18.3 | 21.0 | 23.6 | 26.2 | 31.4 | 41.9 | 52.4 |
| 7 | 3.1 | 6.1 | 9.2 | 12.2 | 15.3 | 18.3 | 21.4 | 24.5 | 27.5 | 30.6 | 36.7 | 48.9 | 61.2 |
| 8 | 3.5 | 7.0 | 10.5 | 14.0 | 17.5 | 21.0 | 24.5 | 28.0 | 31.5 | 35.0 | 42.0 | 55.9 | 69.9 |
| 9 | 3.9 | 7.9 | 11.8 | 15.7 | 19.7 | 23.6 | 27.5 | 31.5 | 35.4 | 39.4 | 47.2 | 63.0 | 78.7 |
| 10 | 4.4 | 8.7 | 13.1 | 17.5 | 21.9 | 26.2 | 30.6 | 35.0 | 39.4 | 43.7 | 52.5 | 70.0 | 87.5 |
| 11 | 4.8 | 9.6 | 14.4 | 19.3 | 24.1 | 28.9 | 33.7 | 38.5 | 43.3 | 48.1 | 57.8 | 77.0 | 96.3 |
| 12 | 5.3 | 10.5 | 15.8 | 21.0 | 26.3 | 31.5 | 36.8 | 42.0 | 47.3 | 52.6 | 63.1 | 84.1 | 105.1 |
| 13 | 5.7 | 11.4 | 17.1 | 22.8 | 28.5 | 34.2 | 39.9 | 45.6 | 51.3 | 57.0 | 68.4 | 91.1 | 113.9 |
| 14 | 6.1 | 12.3 | 18.4 | 24.6 | 30.7 | 36.8 | 43.0 | 49.1 | 55.3 | 61.4 | 73.7 | 98.2 | 122.8 |
| 15 | 6.6 | 13.2 | 19.7 | 26.3 | 32.9 | 39.5 | 46.1 | 52.7 | 59.2 | 65.8 | 79.0 | 105.3 | 131.7 |
| 16 | 7.0 | 14.1 | 21.1 | 28.1 | 35.1 | 42.2 | 49.2 | 56.2 | 63.2 | 70.3 | 84.3 | 112.4 | 140.5 |
| 17 | 7.5 | 14.9 | 22.4 | 29.9 | 37.4 | 44.8 | 52.3 | 59.8 | 67.3 | 74.7 | 89.7 | 119.6 | 149.5 |
| 18 | 7.9 | 15.8 | 23.8 | 31.7 | 39.6 | 47.5 | 55.4 | 63.4 | 71.3 | 79.2 | 95.0 | 126.7 | 158.4 |
| 19 | 8.4 | 16.7 | 25.1 | 33.5 | 41.8 | 50.2 | 58.6 | 66.9 | 75.3 | 83.7 | 100.4 | 133.9 | 167.3 |
| 20 | 8.8 | 17.6 | 26.4 | 35.3 | 44.1 | 52.9 | 61.7 | 70.5 | 79.3 | 88.2 | 105.8 | 141.1 | 176.3 |
| 25 | 11.1 | 22.2 | 33.3 | 44.3 | 55.4 | 66.5 | 77.6 | 88.7 | 99.8 | 110.8 | 133.0 | 177.4 | 221.7 |
| 30 | 13.4 | 26.8 | 40.2 | 53.6 | 67.0 | 80.4 | 93.8 | 107.2 | 120.6 | 134.0 | 160.8 | 214.4 | 267.9 |
| 35 | 15.8 | 31.5 | 47.3 | 63.1 | 78.8 | 94.6 | 110.4 | 126.1 | 141.9 | 157.6 | 189.2 | 252.2 | 315.3 |
| 40 | 18.2 | 36.4 | 54.6 | 72.8 | 91.0 | 109.2 | 127.4 | 145.6 | 163.8 | 182.0 | 218.4 | 291.2 | 364.0 |
| 45 | 20.7 | 41.4 | 62.1 | 82.8 | 103.6 | 124.3 | 145.0 | 165.7 | 186.4 | 207.1 | 248.5 | 331.4 | 414.2 |
| 50 | 23.3 | 46.6 | 69.9 | 93.3 | 116.6 | 139.9 | 163.2 | 186.5 | 209.8 | 233.2 | 279.8 | 373.0 | 466.3 |
| 60 | 28.9 | 57.7 | 86.6 | 115.5 | 144.3 | 173.2 | 202.1 | 230.9 | 259.8 | 288.7 | 346.4 | 461.9 | 577.4 |
| 70 | 35.0 | 70.0 | 105.0 | 140.0 | 175.1 | 210.1 | 245.1 | 280.1 | 315.1 | 350.1 | 420.1 | 560.2 | 700.2 |
| 80 | 42.0 | 83.9 | 125.9 | 167.8 | 209.8 | 251.7 | 293.7 | 335.6 | 377.6 | 419.5 | 503.5 | 671.3 | 839.1 |
| 90 | 50.0 | 100.0 | 150.0 | 200.0 | 250.0 | 300.0 | 350.0 | 400.0 | 450.0 | 500.0 | 600.0 | 800.0 | 1000.0 |
| 100 | 59.6 | 119.2 | 178.8 | 238.4 | 297.9 | 357.5 | 417.1 | 476.7 | 536.3 | 595.9 | 715.1 | 953.4 | 1191.8 |
| 110 | 71.4 | 142.8 | 214.2 | 285.6 | 357.0 | 428.4 | 499.9 | 571.3 | 642.7 | 714.1 | 856.9 | 1142.5 | 1428.1 |
| 120 | 86.6 | 173.2 | 259.8 | 346.4 | 433.0 | 519.6 | 606.2 | 692.8 | 779.4 | 866.0 | 1039.2 | 1385.6 | 1732.1 |

Some radar vendors will use wide vertical beamwidths to eliminate the need to adjust elevation. Although this may work for many locations, it does not always give the desired result. Most vertical beamwidths are less than 25 degrees, which means 12.5 degrees below level (or 0 degrees elevation) and 12.5 degrees is above the level. If a radar has already been purchased, look at its specifications concerning the elevation beamwidth before going on the survey.

Chapter 4 discusses how to use Google Earth as well as a surveyor's transit to find the elevation coverage at a specific site.    For the technology savvy, the surveyor can use software to determine radar coverage.    These are costly programs and can take some time to setup each run.    ESRI offers this capability in some of their mapping software. This software will show all areas the radar has line-of-sight, as well as areas that cannot be covered due to terrain obstruction.

Elevation pattern is most important when it comes to sites where both long range and short range detection is needed.

Figure 5-1 shows the importance for tower height, vertical beamwidth, and desired coverage on the ground for radar.    The upper left picture shows a radar tower that is too high.    The upper right picture is better, but the radar beam still creates a spot on the ground.  These top two figures show what is known as the **spotlight** mode for radars.    This should be avoided because it requires the radar to tilt down on one pass and tilt up on a second pass.   This means it takes twice as long to cover the area of concern.    The bottom picture shows a radar at the correct

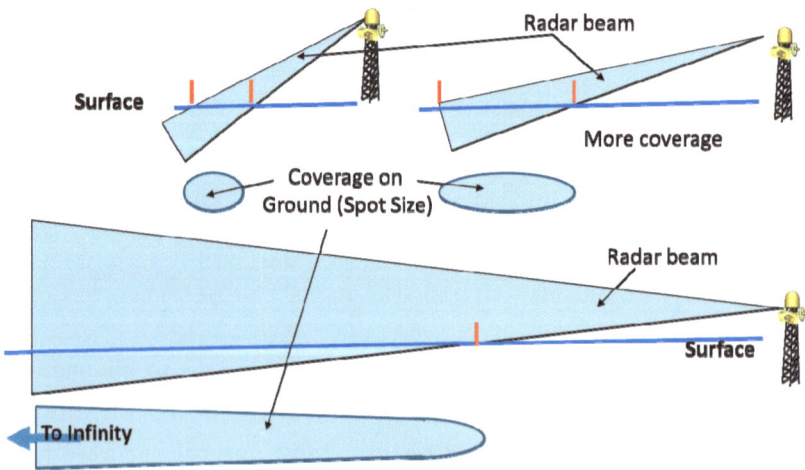

*Figure 5-1. Beam Coverage of Radar for Various Tower Heights.*

height. Its beam shoots off to infinity and does not require two different tilt angles over two passes.

Normally the longer the range of the radar, the smaller the tilt extent it possesses. For instance, the AIMS Fast-Scan radar sold by DMT has the following tilt range of motion with 0 being perfectly level:

- Short Range configuration: -30 to +90 degrees tilt
- Medium Range, Extended Medium Range, and Long Range Configurations: +/- 15 degree tilt.

Many long range radars have no elevation or tilt capability because the beamwidths are normally large enough to cover vertical terrain change.

Remember that the tilt angle listed in the radar's specifications is for the center of the antenna beam. The elevation beamwidth of the antenna normally covers a few degrees above and below the reported tilt angle.

VIP and prison sites with very tall walls or fences should use radars with large tilt extents, such as the short-range AIMS Fast-Scan radar mentioned above. This radar allows the installer to mount the radar on a high tower to see over walls or trees, but still get in close to the tower. This might take several passes with different tilt settings, but short-range radars normally can spin much faster than longer range radars.

Figure 5-2 illustrates the impact of vertical beamwidth. It also shows what not to do when it comes to site security. A common practice seen throughout the world is to minimize cost by placing one longer range radar on a high tower in the center of a compound. This practice is not recommended, since there are no redundancies and it produces large shadow (no coverage) areas around the fence or wall itself. Having multiple shorter range systems on the perimeter of the site would result in far better coverage, a higher probability of detection, and redundancy.

For all installations make sure the radar horizon limitation is not exceeded. The radar horizon is the furthest distance the radar can detect as a result of the curvature of the Earth. Figure 2-4 illustrates the radar horizon.   The radar horizon can set range limitations that are shorter than the radar's published maximum range or the range stated on its user interface.   To increase the radar horizon for any radar, increase its height above the ground.   There are many online calculators that can be used to calculator the radar horizon.   One, in particular, can be found at:

**http://radarproblems.com/calculators/horizon.htm** .

For any online calculator, be sure to enter 0 (zero) for the target height. This provides margin for uneven terrain, which is not considered in most radar horizon calculators.

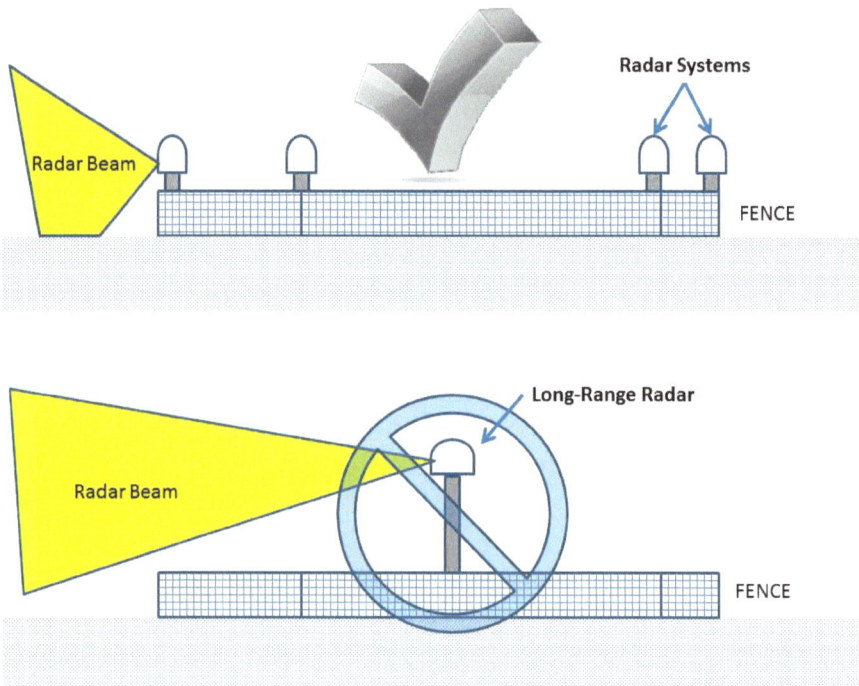

*Figure 5-2.  Placing radar systems lower and closer to the perimeter of the facility results in the best coverage.*

The site surveyor must  to establish the absolute boundaries the customer desires for security.   Often times the customer will set a site perimeter to be small in the hopes of reducing costs.   But once a second radar is required due to obstructions, it is good to let the customer know the perimeter can be expanded if necessary.

When establishing the number of radars needed and their locations, the accepted practice is to have a minimum of 25% overlapping coverage between two or more radars for border security and most other applications. This ensures higher probably of detection – especially for rotating radar systems. For VIP facilities, power plant and prison security, it is good to have overlapping coverage in case there is a security system failure. Examples of this are found in chapter 8.

Also determine if true perimeter protection is needed or would area coverage be better. If the area to be searched by the security sensor is narrow or is close to roads, a line of trees or other walls, then a Fixed-Beam radar is a good solution. Fixed-Beams also work well for corridors and along rivers. If the facility sits on the shoreline of a large body of water or if there are large open areas surrounding the facility, then a rotating radar may be the better choice.

The radar accuracy is not generally an important parameter to the security operator and responders. Most security responders are happy to know the position of the intruder within a hundred meters. It is more useful when it comes to setup and for the site surveyor. The radar's motor encoding resolution and the antenna beamwidth helps determine the azimuth pointing accuracy. The radar's range accuracy is determined by the reflected signal strength of the intruder, the repeatability of the pulse or waveform shape (or bandwidth for FMCW), system clock accuracy and the A/D (analog-to-digital) board. Appendix B has the formula for calculating range accuracy. The positional accuracy is a function of azimuth and pointing accuracy, range accuracy, and the tracker performance.

For the site surveyor, radar accuracy will be most important for VIP, Base and similar applications. As a radar spins, its beam will come close to the wall or fences of these sites. Overshoot or large inaccuracies will result in false or nuisance alarms. Radar's with greater than one beamwidth error in azimuth pointing accuracy will be hard to point during setup. Towers for these radars should be set farther out from walls and fences than Table 5-1 indicates.

Range inaccuracy can cause nuisance alarms in setups like the VIP residence site shown in Figure 4-1.    In this case, siting does help with radar performance. During installation and setup, the radar beams in the direction of the roads should be decreased in range to insure the beams will not detect people on the sidewalks and traffic on the roads. Radars like the MDR, or Motion Detection Radar (a perimeter radar by DMT) are range-gated radars. ***Range-Gated Radar*** divides its beam in range into segments.   (The term "range bin" is sometimes used instead of "range gates.") Most radars sold today for security are range gated systems. For 1.5 km range, the MDR (Motion Detection Radar, by DMT) will break up the beam into 500, 3-meter gates. This means that range can be limited to within 3 meters and allows for installations in Figure 4-1 to be easily secured with no nuisance alarms. Very short range and low-cost beam-breaker radars do have problems since they are not range gated. They rely on reduction of radar power to reduce range.

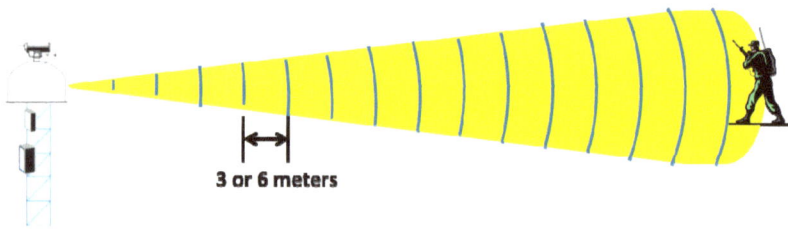

3 or 6 meters

*Figure 5-3.  Many radar systems can break up their range into segments known as <u>Range Gates</u>. These are useful for limiting range in short range applications like VIP, Nuclear Power Plants, Prison, and similar security applications.  The gate size varies from 1 to 100 meters by most security radars.*

Understanding tower installations are important for the site surveyor. Installers at DMT like to orient towers so that the home positions of the radar and cameras are pointing to True North.   Most user displays are arranged with a map that has True North at the top of the page. So alignment and setup becomes easier when the sensors are aimed North.

Before going on the site survey, contact the radar vendor for their mounting bracket solution or the systems integrator if they are responsible. Compare this mounting solution with the tower selected and determine whether there is anything that will impede tower orientation toward True North. If you cannot get this radar or tower information prior to the site survey, be sure to take pictures of the sites selected so that you can make decisions on orientation after the information is obtained.

# 6 Camera Site Surveys

Camera site surveys are thought to be considerably easier than radar site surveys. Although this is true for the most part, the process should not be taken likely. The next few paragraphs revisits key points about each camera technology.

CCTV cameras are generally only short range products and are used primarily for daytime surveillance. Popular dome cameras, which can spin 360 degrees and tilt over a wide range of elevation angles, are often low-light cameras that can see in dim light. But even today's best CCTV cameras cannot see in the pitch blackness of a moonless night.

Figure 6-1 is a scene at night from a CCTV camera. Figure 6-2 is the exact same scene taken moments later with an uncooled thermal camera. Near the center of Figure 6-2, the white hot signature of a human at the distance of 500 meters is clearly visible. Figure 6-3 is the same scene with the same exact thermal camera where items with heat are black in color.

Thermal cameras are essential for many security applications. They are great for nighttime operation, for places with heavy fog, or if there is the potential threat of smoke. One of the detriments to thermal cameras is background heat as seen in the pictures in Figures 6-2 and 6-3. If the person were to walk in front of the building in the background, he/she might go undetected. Many uncooled cameras have only one or

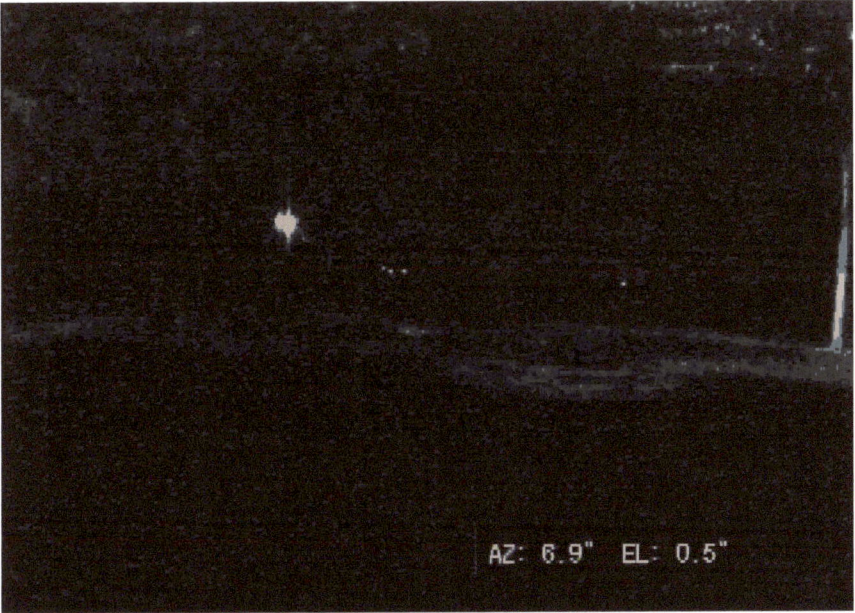

*Figure 6-1. Picture from CCTV on a moonless night. Foreground trees are lit by nearby office lights.*

*Figure 6-2. Picture from uncooled thermal camera of the same scene as in Figure 6-2. All items with heat signatures are white in color.*

AZ: 6.9"  EL: 0.5"

*Figure 6-3. Same scene as in Figure 6-2, but with black hot imaging.*

two field of views (zoom levels) due to their limited range. Cooled cameras are normally supplied with multiple field of views and even continuous field of views. So cooled cameras can overcome this background saturation problem by providing closer looks at the intruder. Thermal cameras cannot see through glass.

Laser-illuminated and LED-illuminated cameras can see through glass. They have longer range performance than uncooled thermal cameras, and are shorter range and lower cost than cooled thermal cameras. There are some camera manufacturers that are building thermal, daylight, and laser illumination on a single pan-tilt unit. At the time of writing, most manufacturers max out at 3000 meters in range for laser-illuminated cameras and 500 to 1000 meters for LED-illuminated cameras.

So from the above discussion, the camera type, mission, and background scene must be considered when siting a camera. Avoid heat signature sources in large parts of the background scene when using thermal cameras. Reconsider using laser-illuminated cameras on the waterfront due to poor fog performance. Or at least plan to use both thermal and laser-illuminated solutions to get the best solution overall.

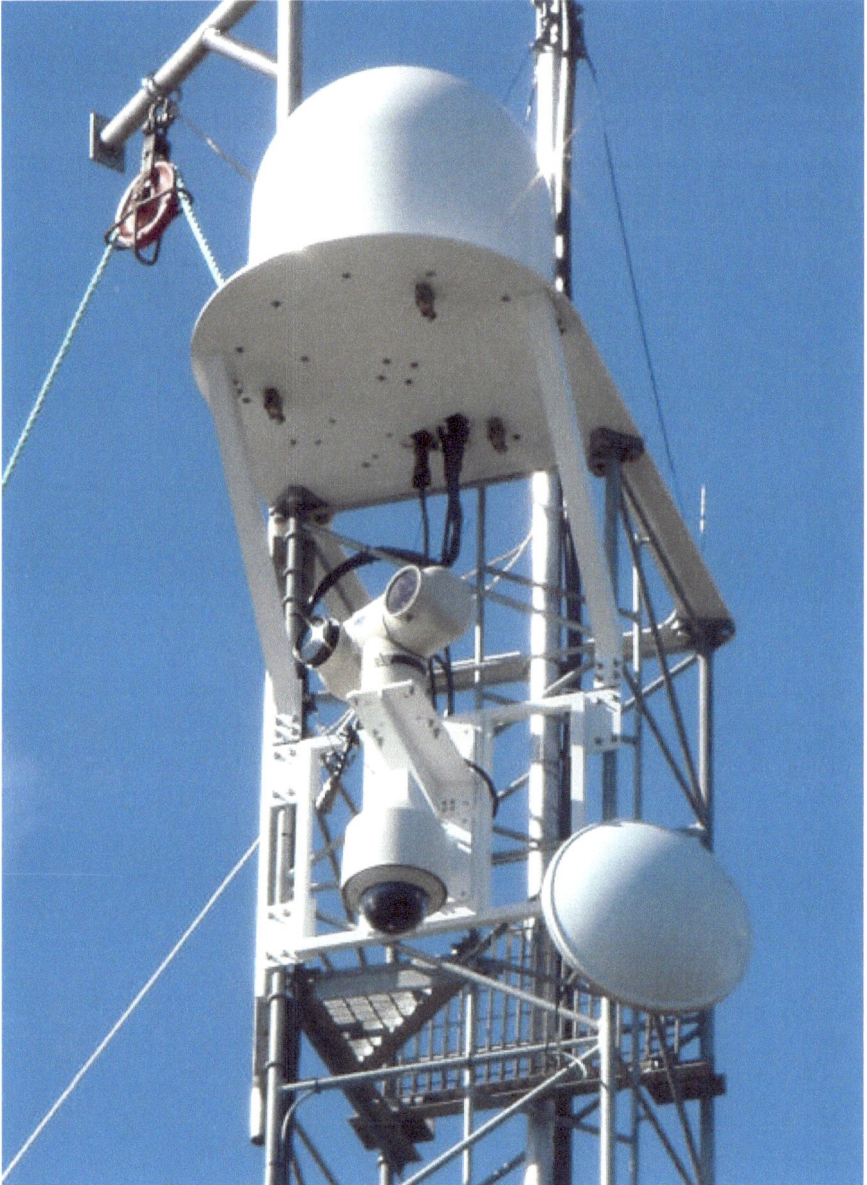

*Figure 6-4. This waterside installation has a combination thermal and daylight camera that is controlled by the radar above it. The lower dome camera is used for tower self-protection. In this arrangement, the radar can be serviced or removed without removing either cameras. And of course, either camera can be maintained or remove easily without disturbing other equipment.*

The mounting of the camera is very important. To ease setup, its most convenient to align the home pin of the camera with the home pin of the radar. The camera can be mounted either on its own bracket or on top of other sensors. Figure 6-4 shows the camera mounted separately from the radar. Figure 6-5 shows the camera mounted on the top of a radome using an adapter plate, and Figure 6-6 and Figure 1-1 show cameras bolted directly to the radome (no adapter plate needed).

*Figure 6-5. The laser-illuminated camera above is shown mounted on the top of a radar. This affords the camera and radar a 360 degree view. Note the metal adapter plate between the radome and the camera base.*

*Figure 6-6. This shows a camera base screwed directly in the top of a radar radome. As with Figure 6-5, this offers the radar and camera with a 360 degree view.*

Although it is nice to have the 360 degrees of unobstructed view offered by mounting the cameras on top of the radars, the site surveyor should consider maintenance. If the radar requires frequent maintenance, it becomes much more difficult to perform that maintenance if the camera must also come down with the radome.

Getting video back to the command center is always a major concern. Running long lengths of coax is not an option. Conversion to digital video has become very popular. This requires high bandwidth lines. Care should be taken to avoid delays, which can sometimes occur with older types of encoders and when the network is overloaded. A good rule of thumb is to assign at least 2 MBPS data rate for the network for each video stream needed. Another popular method is the conversion first to fiber and then back to coax or encoder at the command center.

There has been a large shift to TCP/IP based control of cameras. TCP/IP control is the most practical and easiest means of camera control. These cameras often have a military-style connector or a weatherproof RJ/45 connector for TCP/IP communication.   Older and sometimes alternative methods of controlling cameras are RS-232, RS-422, or RS-485 serial connections.   There are substantially more setup parameters required for control with serial communication.  And because severe limitations in the distance over which serial communication wires can be run, serial-to-fiber converters are often used.   Unless the radar is controlling the camera, it is preferred to purchase cameras with TCP/IP communication options.   Camera control is a low-bandwidth task.   Camera control communication is predominately ASCII text characters or binary.

Take care in the selection of the camera when it comes to video frame rate and power input.   The USA uses NTSC video (30 frames per second) and power with DC voltages, 24 VAC,  or 115 VAC (60 Hertz).   In Europe, the Middle East and many other locations worldwide use PAL video (25 frames per second) and power with DC voltages, 24 VAC, or 220 VAC (50 Hertz).   Mixing monitors and video frame rates can cause undesirable affects or loss of video.   So choose with care.   Many modern video servers have autosensing capabilities for NTSC and PAL, so installation becomes a little easier when these are used.

As with radar, the camera can be limited in range by the Earth's limb.  The **Optical Horizon** can limit the camera from seeing to its maximum range. The optical horizon is shorter than the radar horizon. So when siting a tower with both radar and camera, place the camera higher on the tower than the radar.   The website calculator mentioned earlier for radar horizon calculation also has an optical horizon calculation.  This link is:

*www.radarproblems.com/calculators/horizon.htm*.

As discussed before, enter zero for the intruder height if it is a person.

Vibration is another concern for site surveyor when it comes to cameras. The larger the camera, the more susceptible it will be to wind-driven vibration.   And the larger and heavier the camera, the greater the potential of tower sway from gusting wind.   Check with the camera vendor to see if stabilization or vibration isolation is required.   When vibration is too great, cameras with autofocus will go in and out of focus, which can be annoying to the operator.   Manual-focus cameras will require constant adjustment when there is heavy vibration or constant tower sway, which is also annoying.

Some cameras come with electronic stabilization, which is usually a software-based correction for jitter. Using frame registration, images are aligned on a frame-by-frame basis to remove the vibration or jitter effects. A small amount of the video window is lost in electronic stabilized cameras. Electronic stabilization can also be accomplished with external hardware which performs the same task as the software-based solutions.

When vibration or tower sway is too great and cannot be corrected by electronic stabilization and tower guying, mechanical stabilization is needed. Mechanical stabilization using gyrostabilizers or accelerometers will correct for most types of large and small vibration, as well as, shock and large sudden motion. But mechanical stabilization is expensive. So the site surveyor should study the tower loading and wind specifications to ensure minimal tower motion if mechanical stabilization is not an option.

# 7 Towers and Communication

## *Towers*

Towers are rated for capacity and wind performance versus height. Wind loading is just as important as the amount payload it can support. Every tower manufacturing company supplies detailed specifications with these parameters. The site surveyor should become very familiar with these.

The wind loading on a radar is based on the size and shape of the radar. Radar systems that have a radome (radar cover) have may have larger wind loading values just because they are so much larger than radars without radomes. There are a number of standards for calculating wind loading, including EIA(Electronic Industry Association)-222 and UBC (Unified Building Code) 1997 revision. A good estimate for wind loading is:

$$F = A \ x \ P \ x \ C_d$$

*where*

F = Force (loading) of wind
A = projected area of the radar (square feet)
P = Wind Pressure (PSF) = $0.00256 \ x \ \text{Wind Speed(mph)}^2$
$C_d$ = Drag Coefficient.

$C_d$ for a radome ranges from about 1.2 to 1.5. For a radar without a radome and with flat-panel antennas, $C_d$= 2.

Every country has their own guidelines, procedures and laws for tower installation, which includes the grounding of those towers. So please consult your local tower installation company and electrician as the final authority on any grounding activity.

## Soils and Tower Grounding

Soil type plays a big role in tower grounding. The list below shows in ascending order the electrical resistance of soil. For grounding purposes, soils types of 5 and 6 provide the worst grounding.

1. Wet marshy lands, or soil containing ash
2. Clay, loamy soil, arable land clay
3. Clay & loamy soil mixed with gravel & sand
4. Damp & wet sands
5. Dry sand
6. Gravel & Stones

To improve grounding, chemicals can be added to the ground around the grounding rods. To reduce corrosion, the chemicals should not be in direct contact with the rods. Salt (sodium chloride, magnesium chloride, calcium chloride, sodium nitrate, magnesium sulphate) will improve grounding for poorly conducting soil. It is often recommended that a homogeneous blend of charcoal, salt and sand should surround the rods. Keeping the soil moist will also improve grounding. Both chemical and water treatments need to be replenished over time. Once-a-year replenishment of salts is suggested.

## Ground Rods

Grounding is mandatory for any tower or pole in which sensors will be mounted. Short towers and poles, say less than 20 feet in height, can get by with one very good quality grounding rod. For tall tower installations, it is recommended that at least 3 long grounding rods be used. Typically these rods are arranged outwardly from the tower legs (supports). Three meter (or longer) copper rods are a good solution for grounding rods. The same material should be used for all the rods. These should be placed inside 3 meters from the tower itself. Spacing of the rods should be roughly two times the length of the rod or greater.

In very dry desert soil, the tower may require a grounding cage. These are holes dug 3 x 3 meters wide x 5 meters deep holes for the ground rods. In these holes a copper (or other metal if copper is not available) cage is constructed around the grounding rods. #6 copper stranded cable is often used for a solid ground on the tower itself.

## Measuring Ground Effectiveness

Most radar systems are 50 ohm devices, so grounding measurements should be well below this value. In tower installations completed by DMT, a 5 ohm or lower ground resistance measurement is a good rule of thumb. 10 ohms should be the maximum if at all possible. The installer should always rely on measurement, not just calculation, for tower grounding. This is because temperature and moisture content can greatly influence soil resistance.

Grounding test equipment should be used for measuring the resistance. Many vendors sell good quality testers. The author's installation team uses the Extech Instruments Model 382357 ground resistance clamp-on tester, which comes with calibration hardware to insure a quality measurement.

## *Mounting Equipment*

Radar and camera systems are normally mounted on wall or building mounts, poles, or towers. The Rohn 45G tower (left in Figure 7-1) is the most widely used tower for small to medium sized radar systems and cameras. Larger radar systems employ the Rohn 65G or even larger tower solutions (see Figure 6-4).

Figure 7-2 shows tower brackets built by DMT for the Rohn 45G tower. For pole-mounted installations, the diameter of the pole increases with height and sensor weight. Figure 7-3 shows a pole mount bracket. This particular one fits an 8-inch, Schedule 80 pipe.

Whether pole or tower, all radar systems need these structures to be firmly planted in the ground or cement. Wind can play havoc to non-Doppler radars and the autofocus of many cameras. Winds can cause the towers to sway. Be sure to abide by the wind rating and load requirements of the tower. Whenever in doubt add additional stabilization to towers, such as guy wires. Guy wires extend far from the tower base, so substantial clearance around the tower may be needed.

***Make sure towers are level!*** Towers are notoriously installed by the good-enough-principle. (That looks good enough to me!) This may be fine for cellular installations, but not radar and

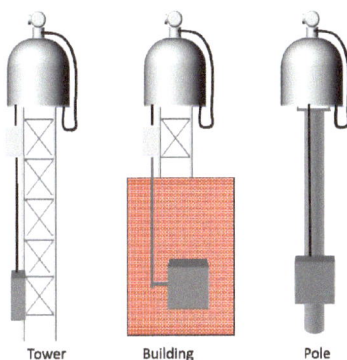

*Figure 7-1. The primary mounts for radar are tower, building and pole mounts.*

cameras. A five degree error (which is not uncommon) is enough to ruin performance of many radar systems – especially long-range systems with tilt (elevation) motors. And do not rely on guy wires to level a tower. Guy wires stretch over time. Some radars, like DMT's AIMS radars come with built-in digital compasses. These compasses monitor bearing, roll and pitch. They are used in the AIMS to compensate for tower errors. But most radar vendors do not include these. Specify a level tower or pole to less than half the radar's beamwidth.

*Figure 7-2. Rohn tower mounts.*

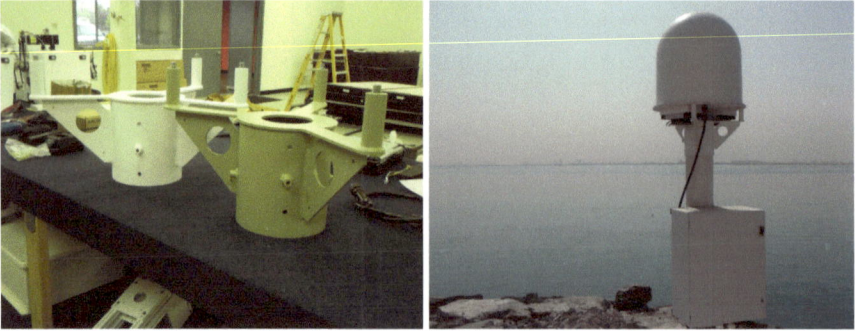

*Figure 7-3. Pole mount shown above. This pole mount is for a Schedule 80, 8-inch diameter pipe. Pipe mounts are a very effective and low-cost solution for mounting. They are easy to level.*

Since radar systems radiate radio waves, it should go without saying that mounting a radar inside a tower is unacceptable.   If you examine a radio and cellular towers, you will notice that the transceivers and antennas are mounted to the outside of the tower using a bracket or on top of it. Most security radar systems  are much higher in frequency than cellular and radio frequencies, so they too should be mounted on the outside or on top of the tower.   Trying to shoot the radar beam of energy through a tower can shorten its range, cause false alarms, and even damage the radar.

When siting a tower of pole, be sure to leave sufficient room for cranes or lifts to maneuver.     Perfect placement for sensor for performance reasons should always be weighed against ease of access.  A tower that is too close to walls, shorelines, roads, or other obstructions may be lead to poor installation, breakage and a lack of maintenance.   Figure 7-4 shows an installation of a DMT radar and a Vumii camera.   This site was well planned for installation, which was completely quickly.

Cables should be properly managed and access enclosures should be out of reach from the ground.   AC power to the top of the tower should be avoided.   It should be converted to DC power at the lowest point on the tower that cannot be reached by a person on the ground.   A fence

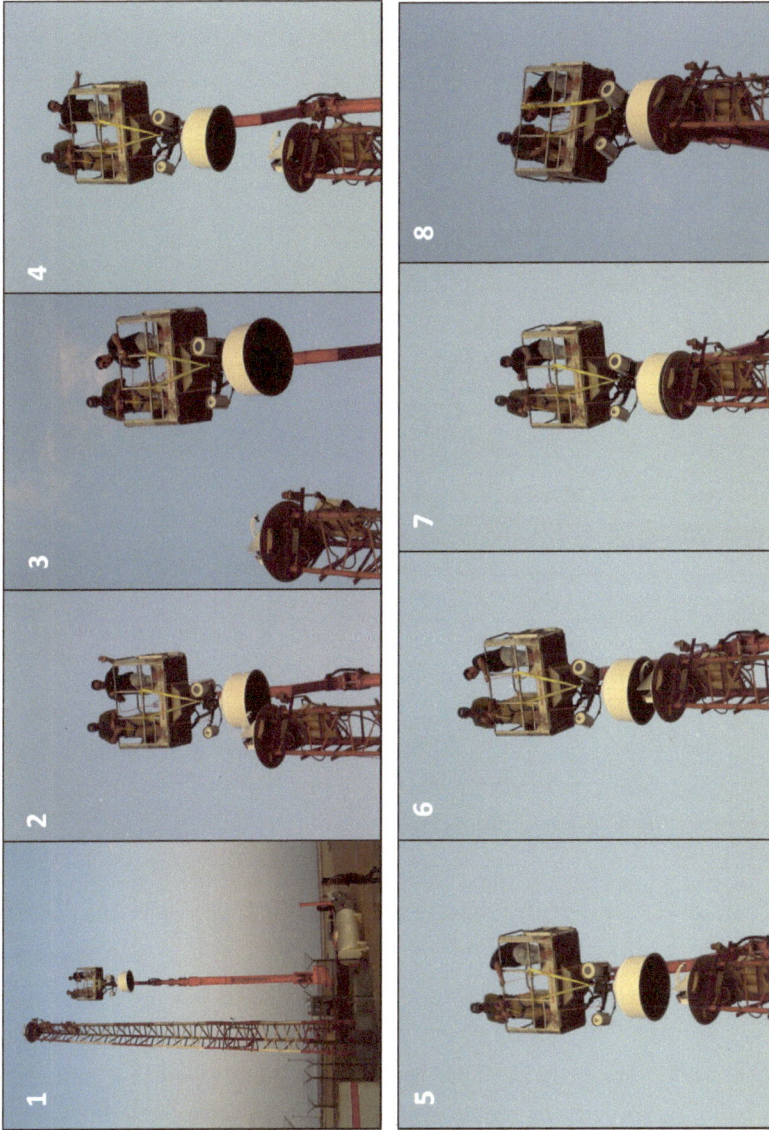

*Figure 7-4. This is a sequence of pictures showing an installation of a radar and camera on a tall tower. There was plenty of room for the crane to maneuver safely.*

around the base of the tower is not required, but does act as a deterrent to pranksters and vandals.

## *Communication*

Communication is an extremely important element of security. It doesn't matter if the sensors are the best in the world if the communications are not reliable and sufficient. Communication methods and equipment cannot possibly be covered adequately in this book, but a few key technologies will be discussed.

Long runs of coax cable and serial communication are rarely seen in security installations today. Most communication for today's security solutions require high bandwidths. Therefore, one or more of the following technologies are usually employed:

- CAT5e or CAT6 runs,
- fiber optics,
- wireless or microwave links,
- cellular networks,
- satellite communication.

Communications over CAT 5e or CAT 6 copper wire are for short runs of 200 feet or less. Many times this type of connection and communication feeds back to a fiber hub or network switch.

Fiber optics come as either multimode (or MM Fiber) or single-mode (SM Fiber). Multimode has become more prevalent in the USA and Europe, whereas, single mode remains dominate in the Middle East. Multimode fiber is good up to about two kilometers for high bandwidth communications. Single-mode can go thousands of kilometers and still maintain high bandwidth. Multimode fiber is far less expensive to purchase and install. If runs of multimode fiber are greater than two kilometers, the surveyor should site the location for a repeater, which requires power and an all-weather enclosure.

When long network or fiber optic runs are not possible or practical, the integrator often turns to wireless links. There are many choices when it comes to this. The most common frequencies selected range from 2 to 10 GHz. These systems are all line-of-sight communications, which means there can be nothing between the pair of antennas that are used for this type of communication. The lower the frequency, the better the performance in all weather. Unfortunately, the lower the frequency the bigger the antenna must be. Dish sizes for longer range wireless communications can easily reach up to 2 feet in diameter. 2.4 to 5.6 GHz are the most commonly used frequencies for wireless data communication. It is now possible to get dual-frequency antennas and wireless modems that sense the environment and chooses the best frequency for maximum throughput. In the USA, 4.9 GHz has been added as an emergency and public service band. Although licenses are required for this, it is a relatively quiet and unused band for government projects.

Decent video streams need at least 2 MBPS data links. To facilitate reduction in performance due to weather, alignment, vibration and aging of cable over time it is good to plan some spare bandwidth. So for a 10 MBPS line, a maximum of four video streams should be planned.

Another option for video is to use video links. These links carry video and audio over microwave links. These links were offered years ago at low frequencies, such as 900 MHz. But now these are sold predominately at 5.8 GHz. They do not have encoders in them like video servers; rather, they convert video to modulated RF of about 3 GHz bandwidth. They are a good option when the microwave links are overburdened or as a backup solution.

Cellular communication is a good option as a backup system. Some security officers and facilities hesitate to use it for the primary means of communication. This is because phone companies manage the network's bandwidth. Bandwidth can be too variable during high usage and the service is susceptible to severe weather outages. It also requires extended service plans, which can be pricey for video. A dedicated service can be purchased, but it is even more expensive. However, this author has seen it used effectively by some facilities as the primary communications method, with slower and lower bandwidth communication sources as backup. And it is also used often for mobile security applications. DMT has successfully used cellular hotspots on its demonstration vehicles to pass video and radar data back to a fixed command station.

Satellite communications are sometimes the only choice for very remote facilities. Satellite communications, as of this writing, are still relatively low bandwidth compared to some the options mentioned. The recurring cost for satellite communications is very expensive. It is often used for mobile security applications.

An important point for the site surveyor is that there should be primary and backup communication planned for all tower locations. The infrastructure required (including cable runs), and antenna line-of-sight issues for communications should be evaluated during the site survey.

# 8 Example Installations

## *Airports, Prisons, and VIP facilities*

Airports are well lit on the inside of their fence line or walls. So inside the facility can be monitored using daylight cameras. Thermal cameras offer improved surveillance in the event of smoke and fog. The outside surrounding areas are usually not lit well at night. So thermal cameras on the boundaries are necessary for airport perimeter security.

Laser-illuminated cameras may not be permitted or desired in many airports. (The fact that laser-illuminated cameras are not permitted or desired may not be a logical reason; rather, it could be simply a broad reaching regulation against lasers of any kind, or a misconception of the dangers of the lasers.) LED-illuminated cameras offer an alternative to thermal and laser-illuminated cameras. They do not normally face the regulatory problems or suffer from the worry of eye injury experienced by laser-illuminated camera installations. However, LED-illuminated cameras cannot see well through fog and smoke, so combining them with thermal camera solutions makes for a good security solution.

Cameras are often the only method of security used for airports. This approach to security results in substantial dedicated manpower. Automated solutions with combinations of radar and other security solutions help reduce manpower and perform better.

Airports, prisons and VIP sites are normally well groomed. The sites themselves are graded flat and have boundary areas that are cleared several hundred meters at a minimum. Short and medium range radar systems are perfect for this application. Terrain following is not required in most of these sites, so elevation motors are not required. Doppler radar systems are not required, but will be useful if there are tall grasses, bushes, or lots of clutter. These sites are ideal for fast spinning radars, such as FMCW radars. Because the flat areas around the perimeter, the radar should be fast spinning.

Airplane signatures are very large and the fence runs are immense. These large objects present a larger signature in the horizontal plane, so a vertically polarized radar should provide better detection of humans and be less affected from saturation effects from fences and the airplanes themselves.

## VIP facilities and Prisons

Prisons and the elaborate dwellings of the rich may seem to have nothing in common with each other. In the security world, however, they are very similar. Each of these facilities are large areas surrounded by fences or walls or both. Each of the facilities are like small self-contained cities. They have workers, places to eat, places to sleep, trash collection, transportation, and their own security. Each tightly control their portals, and the people that pass through those portals.

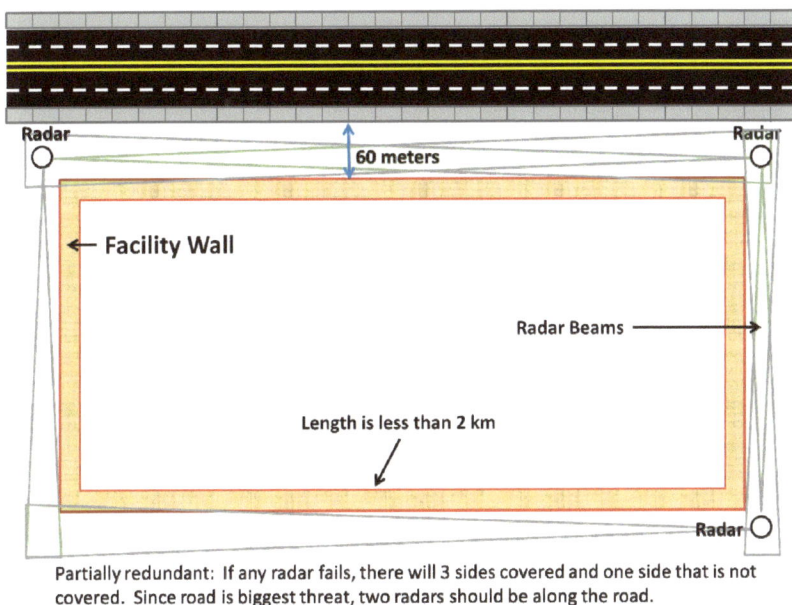

Partially redundant: If any radar fails, there will 3 sides covered and one side that is not covered. Since road is biggest threat, two radars should be along the road.

*Figure 8-1. This is an example coverage for a facility in a more populated area. This is a cost effective solution that has some redundancy. Having a radar on all corners would be fully redundant.*

There are two types of these facilities. Some are large, remote facilities and resemble airports in terms of security. Those can be protected using the same methods used by airports. The second type of facility is in more populated locations. Roads or sidewalks might be nearby. Figure 8-1 is an example of a cost effective, yet partially redundant security solution.

Figure 8-2 is an irregular shaped facility. In this case, the radar is limited to 1.5 km human detection range. Five radars are required for the absolute minimum level of security. In Figure 8-1 and 8-2, uncooled thermal cameras and LED-illuminated cameras are more than adequate for video protection of the facility.

## Border

Border security varies greatly from country to country. For heavy forested borders, unattended ground sensors, beam breakers and

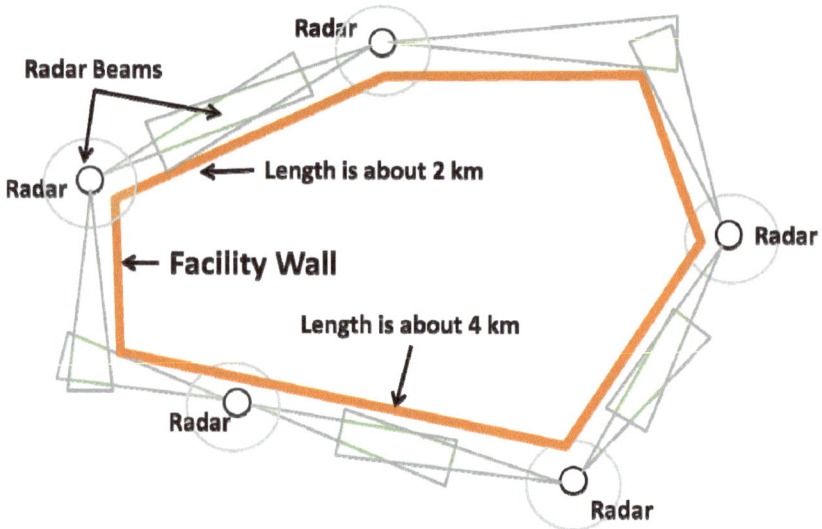

No redundancy: If any radar fails, there will be up to 2 fully exposed walls. 7 radars are required for 100% redundancy.

*Figure 8-2. This palace is minimally secured. It has just enough radar beams to permit full coverage, but not enough for redundancy.*

cameras offer the best protection. Fences slow down intruders, but have not been found to stop the flow of illegal intrusion. Fences also cannot detect intruders, so they are often combined with vibration, fiber optic, or electric charge sensors. Since false alarms may be higher than desired for these types of sensors (especially in inclement weather), camera systems should also be added.

For large open borders, radar is the fast search solution of choice. Unfortunately, most government's choose to employ only long range radars. This approach is not so bad for areas that are flat and open. But often the government chooses to buy these for their entire border. It's a costly and ineffective approach. First of all, what is gained by looking 10 to 20 km over the border? Many roads run parallel to borders and many towns are along borders. So there will be a lot of detections of people with no intent on crossing. Secondly, the longer range radar and camera systems will not detect 100% of people intent on crossing the border.

With only one vantage point, the radar and camera will not have line of sight to all intruders if they are truly intent on crossing.

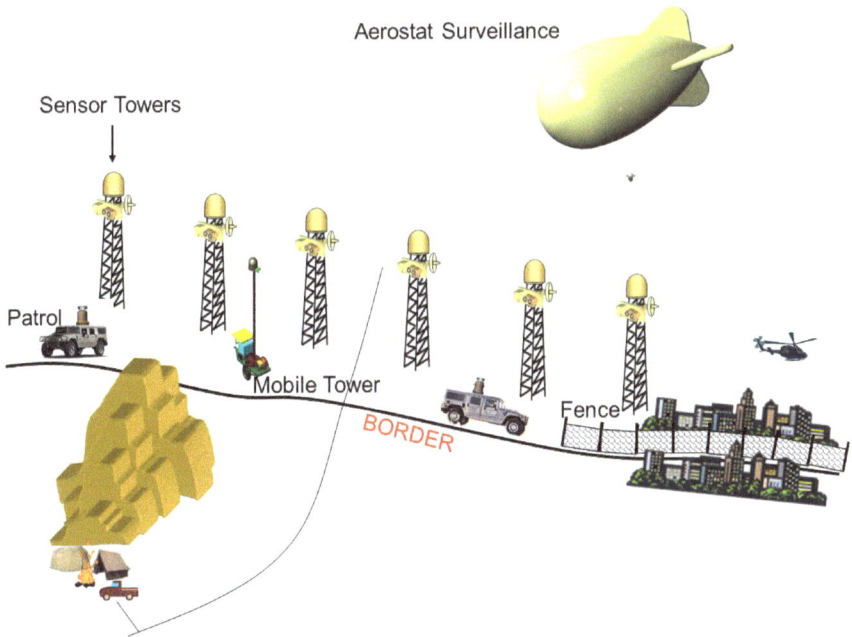

*Figure 8-3. The best border defense is a layered sensor and physical security approach.*

The better choice is to use more mid-range radars with substantial overlapping coverage. This gives higher probability of detection, redundancy, faster detection, and multiple viewpoints. The larger towers and higher power draw required for long-range radars add tremendously to cost. In cost trade studies conducted by DMT, it was found that 5 km radars purchased at a ratio of 3 to 1 to long-range (>= 12 km) radars along a border is a lower cost and better performance solution for borders. There are many more mid-range radar vendors to choose from than long-range vendors. The mid-range thermal cameras coupled with the mid-range radar is the ideal solution.

A multi-sensor approach has been long believed to be the best solution for border security.  Combinations of sustained airborne surveillance (such as that from aerostats, helicopter, small plane, and UAV's), fixed ground surveillance radars and thermal cameras, unattended ground sensors, and vehicle mounted security make borders much more secure than a single method approach.

A newer idea for border security is the use of multiple, overlapping radars systems.  This approach requires two different types of radars with different mission roles.  A fast spinning search radar scans the horizon for potential intruders, and then hands off its tracks to a more sophisticated tracking or interrogation radar.

The search radar can be a high-powered marine radar or FMCW with long detection range.  Although these experience much higher false alarm and nuisance alarm rates, they spin fast and can cover the search area quickly.  This radar acts to cue a Doppler radar with built-in tracker.  Doppler radars are much more effective in removing clutter and false targets than any other technology that may be employed.  By not having to search, the Doppler radar can integrate pulses longer and thereby provide extreme precision.  Since higher integration results in higher average power, pulsewidths can be shorter and offer higher resolution in range.  One long-range search radar can operate with 2 to 4 tracking Doppler radars.  This is an ideal arrangement for radar surveillance and the cost would not be significantly higher.

### *Oil and Gas Sites and Power Plants*

Radar and thermal cameras are a must of oil drilling, processing and storage sites.  The same is said for gas sites, including LNG (Liquefied Natural Gas) and LP (Liquefied Petroleum) gas storage sites.  Larger power plants also use many of the same practices in safeguarding their facilities.

Figures 8-4 to 8-6 are example LNG sites. Figure 8-4 should be used when budgets are tight. If there is a failure in a system, coverage for 50% of the plant is lost.

Figure 8-5 is the preferred arrangement for very remote sites. It has full redundancy. In fact, two radars and cameras could remained powered down to act as permanent spares for any failures of the primary systems.

Figure 8-6 is the preferred solution for most sites—especially for sites that are not extremely remote. For oil drilling or LNG storage sites, three radar and camera systems is a good rule of thumb. If the radar near the gate fails, the system reverts to the minimum configuration (shown in Figure 8-4). And the radar on the right in Figure 8-6 will still be able to detect vehicles or people approaching the main gate.

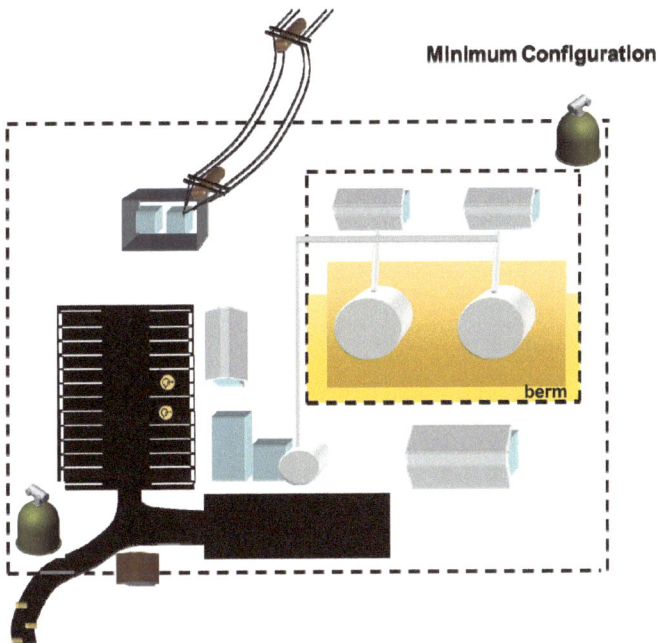

*Figure 8-4. Plan view of a LNG facility. This is a minimum security configuration for a LNG facility. The purpose of this configuration is to protect from outside intrusion to the fence (dashed line).*

87

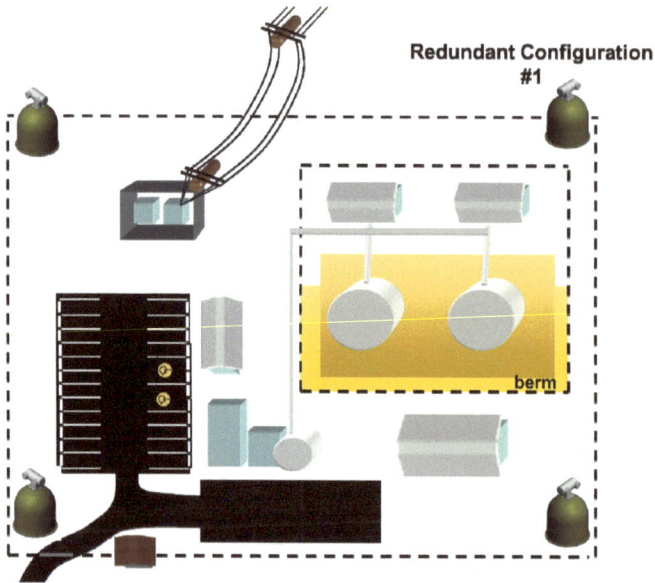

*Figure 8-5. This configuration has 100% redundancy and has the fastest intruder scan rate (update rate).*

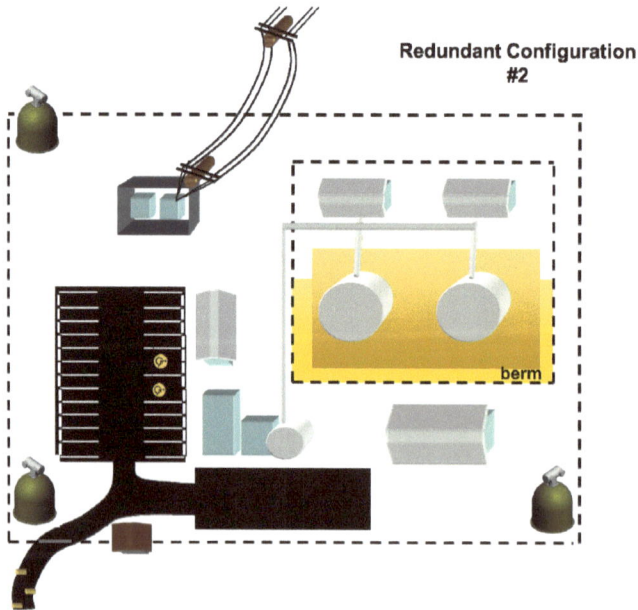

*Figure 8-6. This is solution would be the preferred solution for many sites. It is more cost effective than Figure 8-5, but offers more redundancy and faster updates than Figure 8-4.*

In Figures 8-4 to 8-6, the radars can also be used for the interior security of the facility. EXTREME CARE SHOULD BE TAKEN FOR IN SCANNING AROUND LNG AND LP FACILITIES. The storage tanks shown in the figures are metal. If the beam of a radar spins by the tank, the beam will be reflected away from the tank.

In one LNG facility in the USA, DMT installed a radar near a large LNG tank. The scan was unknowingly set to pass too close the tank. The radar beam reflected off the tank and into the traffic of a nearby freeway. Since the radar didn't know is was being reflected, the car tracks were be plotted inside the LNG facility just beyond the tank. We scratched our heads for a while before the mysterious source of alarms was discovered.

LP gas storage facilities are usually laid out in columns of long cylindrical tanks. Just like LNG facilities, they will be surrounded by a fence. These tanks can also play havoc to a radar – it is like a sea of metal. A set of fixed-beam radar systems around the tanks is a far better, low-false-alarm method of protection of these plants.

An finally, note the buildings in Figure 8-6. In many industrial facilities in the USA, the buildings are often metal – including the roof. Ribbed metal rooftops present large radar signatures. So radar systems installed near these buildings should be mounted significantly higher or lower than the roofline to avoid the large reflections from the roofs of these buildings.

## *River*

Riverside security is a challenging task. Flowing water at varying rates combined with precipitation and shoreline windblown clutter makes operation difficult. Add the requirement of low false alarm rates and it becomes almost impossible for many types of radar systems.

Video motion detection just doesn't work in this kind of environment. Also, if someone is wearing a neoprene wetsuit their signature is undetectable by thermal cameras because their outside temperature is the same as the river water.  So radar is required for rapid, all-weather detection of intruders on and in the water.

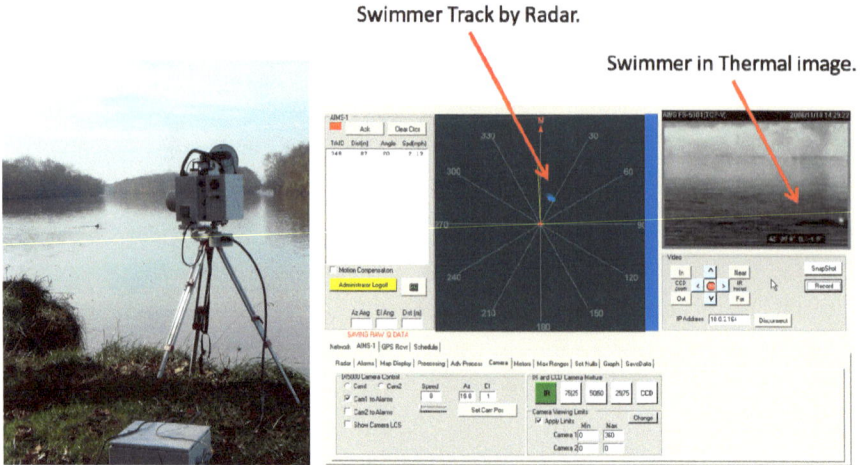

Figure 8-7. Swimmers being detected on a river.  Radar is pointing thermal camera automatically.  Swimmer has wetsuit on, but face is visible as bright white. Rotating radars, such as this one (which has the radome off), is great for wide rivers such as the Potomac (pictured), Delaware, Mississippi, Missouri, Colorado, and similar.

Doppler X-Band radar is excellent for riverside operation (see Figures 8-7, 8-8, 8-9).  Removal of the velocities generated by the flowing water is accomplished using:

- Doppler notches centered on 0 Hertz and that are automatically widened when waters flow faster and with more force;
- Doppler notches placed on frequency bins associated with the flowing water radial velocity component (i.e., notch frequency bins to one side of 0 Hertz);

At DMT we found that our X-Band Doppler radars worked even better in the water environment than in the land environment scenarios.

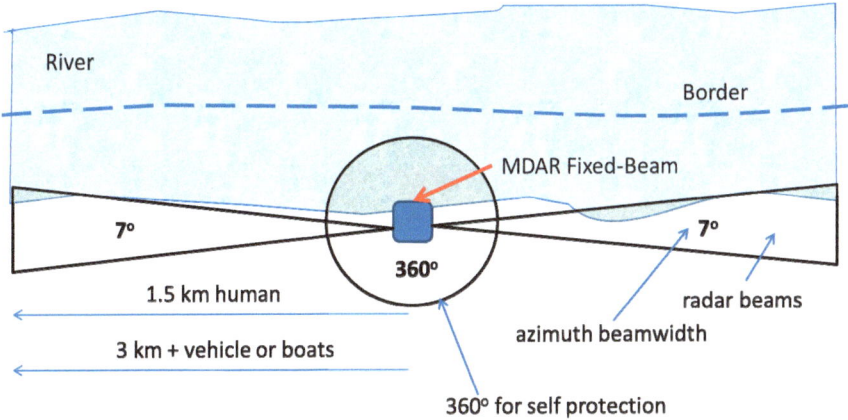

*Figure 8-8. The MDAR is a Fixed-Beam, X-Band, Doppler radar that works well for waterside detection. In this figure, it is protecting shoreline access.*

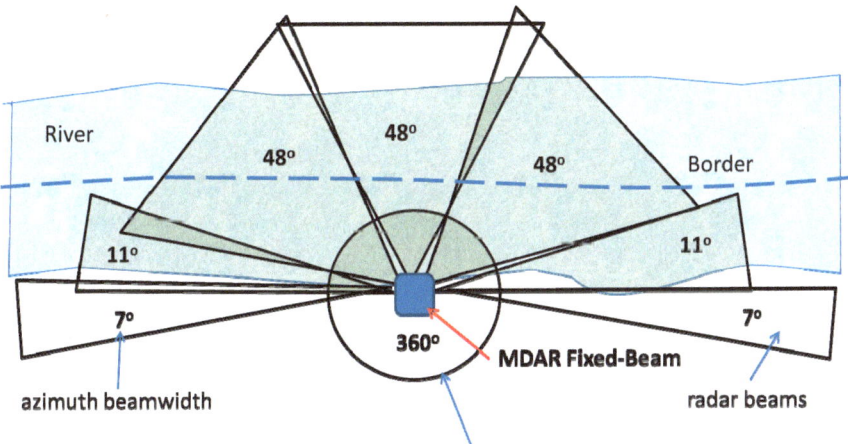

*Figure 8-9. Here the MDAR is used to protect the shoreline and monitor the river itself. Shorelines are protected well; however, the systems will not monitor entire sections (river and shoreline) for very wide rivers.*

Marine radars combined with clutter subtraction and tracking also work well for this type of application. Marine radar systems spin fast, which is good for detection of jet skis and speed boats.

What should one look for in a good riverside radar? The selection should have at least five of the following six characteristics:

- C-Band to X-Band frequencies offer the best compromise – their wavelength keeps the radar system a manageable size, sea clutter rejection is good, and performance in all types of weather is good. Be careful, however, of the frequencies 9.35 to 9.45 GHz for the radar you select. This is the marine radar band and there will be a potential for interference, unless the radar is a Doppler radar. If the restricted band above cannot be avoided, make sure the radar is still capable of operating in the presence of other shipboard radars. Ku-Band radar solutions are totally inappropriate for these applications.

- H-Polarization has been widely known to be a better sea clutter rejection polarization than V-Pol. Since boats typically have longer dimensions in the horizontal plane, it makes for better detectability of water craft.

- Doppler radar systems offer superior sea clutter rejection. The same is true for river clutter rejection.

- A very good tracker is essential. If this is available, then FMCW radars can be used as long as they are C-Band to X-Band emitters. By plotting only mature tracks, the radar may operate without the numerous false alarms generated from poor clutter rejection.

- Fast update rates are required for riverside radars. Objects that navigate across the river can do so quickly. Speed boats moving down river can exceed the speeds of land vehicles. So a fast spinning or fast scanning radar is required. If the radar cannot maintain at least a twenty degree or greater spin rate, use overlapping radar coverage from two radars to maintain a high probability of detection.

- Ability to blank or null buoys that bob in the water. Buoys in bays and harbors move back and forth with tidal changes over the course of one day. Buoys in rivers tend to remain at a constant location. Current keeps their tethers stretched along the direction of the river. In turbid waters, the buoy can move substantially up and down. There may also be some lateral motion, but again not as much is seen in ocean or harbor buoys.

A null or blanking zone placed around the buoy will remove this nuisance without dramatically affecting overall performance.

What is the best camera to use along the riverside? Thermal cameras are a must. They allow visual detection of people and boats. They also see through some fog and smoke. As was mentioned before, if a person is wearing a wetsuit they will be virtually undetectable. Laser-illuminated cameras can "see" the person emerging from the water in a wetsuit, but that is in good weather. Fog can severely limit the performance of the laser-illuminated camera. So the best camera solution would be a combination camera set. A thermal camera combined with daylight camera and laser or LED illuminator would be the better choice.

Before leaving the subject of river security, we should briefly mentioned bridge security. Bridges are key to the transportation of goods across waterways. All major bridges should be secured with at least video camera security. This should be done above and below the bridge deck. Dome cameras, such as Pelco's, Bosche, Panasonic, Sony, or any other similar outdoor camera should be the minimum. Radar systems enable advance warning for potential intruders to the pylons. But it should be noted that these support structures are a favorite of fishermen, so any detection hardware (radar or otherwise) will have to contend with the sport fishermen.

And a final discussion about power and communications is warranted. UPS (Uninterruptable Power Supplies) are always recommended at any installation. But the riverside security offers a problem not experienced by other land sites – floods. When planning for the UPS system, be sure to install it above the highest known flood elevation for that river. This may be difficult for two reasons:

- A core element of an UPS is the batteries. Batteries are heavy and it is not uncommon for the UPS system to be 400 to 800 pounds when loaded with the batteries. This can be a challenge to mount above ground level if the UPS is nearby the site.

- Depending on the size and its location, flooded rivers can take days to even a week to drop to normal levels. Power may be disrupted the entire time. Sizing a UPS for days of operation will make it enormous and expensive. A more reasonable UPS approach might be to size it for 2 to 4 hours of power and add a backup generator. See if there is an ability to place the radar and cameras into low-power consumption modes to conserve battery life. For instance, cooled thermal cameras pull more current. So doing without thermal imaging for a while may conserve power consumption. Sometimes backup generators at higher, nearby elevations are used in these situations. A good power plan should let the customer know what to do and how long they can operate in case of power outages.

One last point to make on river security. If at all possible, place systems on both sides of the river. This reduces the height requirement of each tower. Be sure to alternate the placement on each side of the river if it is an available option and have at least 25% overlap in coverage.

## Shipboard

It may seem unusual for some to cover shipboard security in this book because ships are, of course, not stationary. However pirating and terrorist events like the USS Cole have resulted in increased security measures for all large ships when docked.

Besides military vessels, large ships that should have security include VIP yachts, tankers, cargo ships, and cruise ships to name a few. Security solutions are being offered worldwide, but as of this writing few have been implemented.

At a minimum, there should be good, long-range communication systems on all of the vessels mentioned above. Some cruise ships have added some anti-pirate technology, such as LRAD (Long Range Acoustic Device). The LRAD emits directional sound at extremely loud levels. People in the path of the sound ways are often shocked by suddenness of it. One type of sound it emits can make people nauseous. It is considered a non-lethal deterrent.

Tankers have also employed water cannons to help drive terrorist from approaching and boarding the vessel. This combined with an LRAD has shown to be effective if there is advanced warning of approaching threats.

Shipboard radar used for navigation can give some advance warning. However, a better approach is to use security radar and thermal cameras for advance warning. Doppler security radars work great for this application. They can help remove sea clutter to see harder to detect objects. Thermal cameras provide detection of heat signatures and can see at night and through some levels of smoke. LED-illuminated cameras can also be used in good weather conditions. These will allow the operator to see inside the bridge or cabin of the approaching vessel.

Figures 8-10 and 8-11 represent a possible security configuration for a tanker. In addition to LRAD's, water cannons, radar and stabilized thermal cameras, security underwater is provided by a hull-mounted sonar. These systems should be tied together so that the LRAD, thermal cameras and water cannons are directed at potential threats detected by the radar. Firing of the LRAD and water cannon systems should be manually enabled to prevent accidental harm to non-threats.

Other items to compliment these sensors would be a fully stocked safe room (with isolated air shafts that are difficult to reach or find by the intruders), smoke grenades, firearms, a satellite phone, first aid kit, and a flare gun. These items would have to remain locked up and tankers may not wish to carry firearms and smoke grenades given the cargo present. The author has even seen UAV (unmanned airborne vehicles) proposed to give the captain and his crew even further reach and vision from the ship.

It is understandable that ship owners may not wish to go to this level of security due to the expense. But pirate attacks worldwide have cost companies substantial amounts in losses. A bare minimum would be the

# Eddie R Hughes

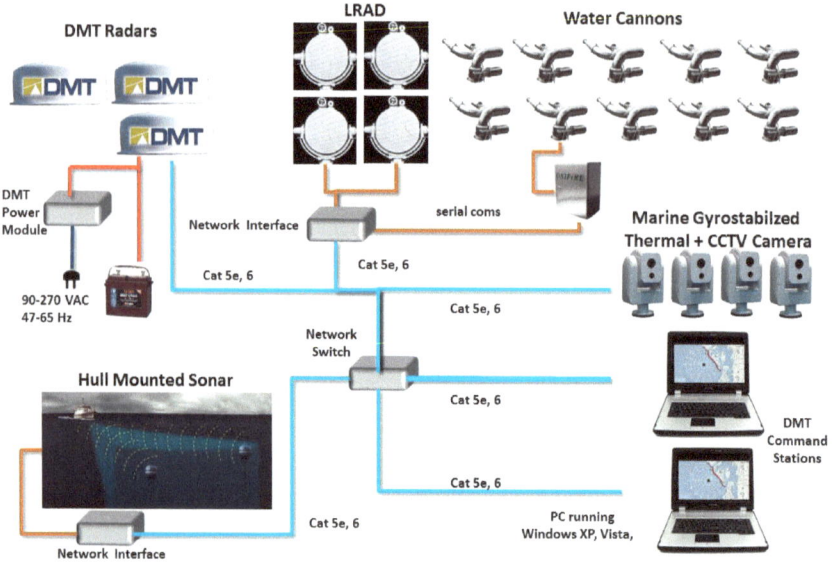

*Figure 8-10. This is a proposed security solution for a tanker. This help protect from threats above and below water. LRAD and water cannons are used for repealing approaching intruders.*

long-range communications, security radar, and cameras. This would give the crew advanced warning and time to notify someone for help.

If a single radar is used for ship security, it is good to put it in a location near the center of the ship. Minimize obstructions to the radar. If budgets permit, two radars located near the bow and the stern would be better. And for extremely large tankers, three radars is the best solution.

### Bay, Port and Shoreline Protection

Unlike riverside security, bays, ports and shorelines have saltwater or brackish (saltwater and freshwater combined). Saltwater is more reflective to RF energy, so when seas are glassy or smooth, RF energy reflects forward. When seas are rough, more energy reflects back to the radar and makes it more difficult to detect objects as well as reducing detection range.

96

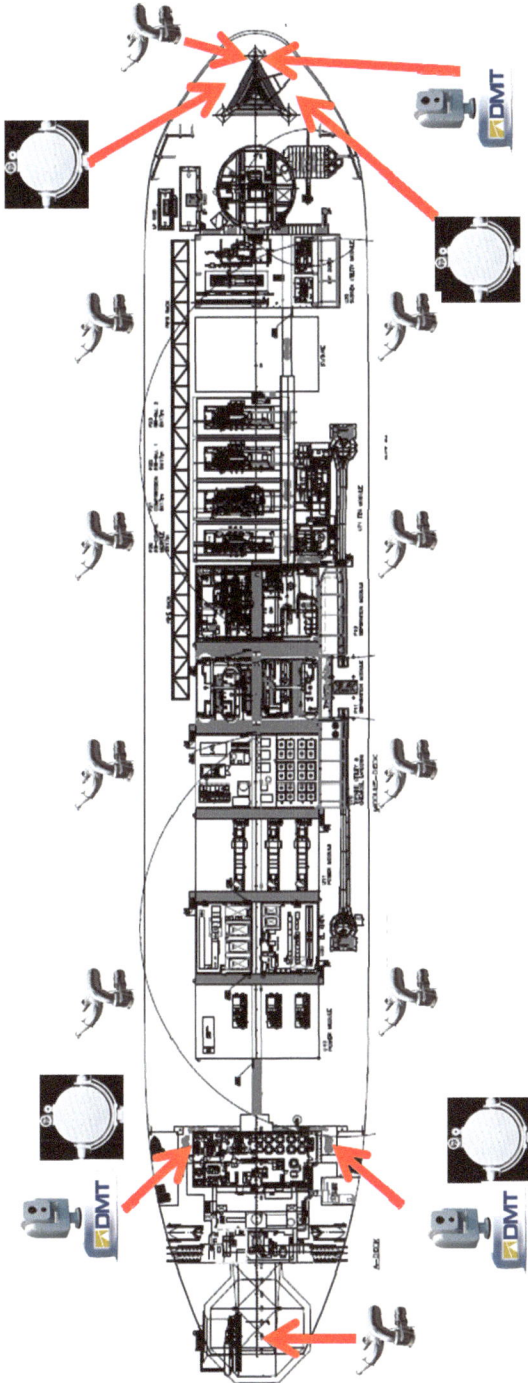

*Figure 8-11    This is the plan view of a tanker, which uses the items shown in Figure 8-10 for security.*

Figure 8-12 shows a facility on the North Sea.  The water here is seldom glassy, so a Pulsed Doppler radar was installed with thermal and CCTV cameras.   One CCTV camera provided views of the base of the tower. This installed security system covers a very large area.   As long as there is unobstructed views, this tower can protect a bay or port that is 3 km in diameter.  Two would be necessary to have the required redundancy and good views on both sides of large ships entering the bay.

The best radars for bays, ports and shorelines are X-Band Pulsed-Doppler and FMCW radars.    Both need good trackers capable of over 50 simultaneous tracks.

Because of fog, thermal cameras work better than laser-illuminated or LED-illuminated cameras for these type of environments.    If budgets permit, a laser-illuminated camera would be a good addition to view inside boat cabins during good weather.    Remember thermal cameras can't see inside windows – especially tinted windows.

### *Lakes and Dams*

Lakes, and especially those with Dams, differ from both shoreline and riverside applications.   Unless we are talking about the Great Lakes or Lake Okeechobee, most lakes do not have nearly the wave conditions seen at seaside installations.  The salinity is non-existent, so reflectivity from the surface is less than the ocean.   So in general, they are easier to protect by radar.

*Figure 8-12.  This tower has a radar and thermal  camera.   This configuration is protecting a several kilometer radius of beachfront.*

If the lake has a dam, the security sensors are often mounted on or near it. This is because there is often structures for mounting towers, power, and communications present. Spillway dams, like the one shown in Figure 8-13, are a little more challenging. It has moving water like a river and then larger areas with virtually no current. So the clutter conditions can be quite varied even within the same sweep when water is flowing over the dam. Many times there are floating lines in front of the dam to catch debris and prevent boaters from getting too close. These lines will move with water flow, so radar software filters must accommodate this.

X-Band FMCW and Pulsed-Doppler radars are great for lakes and dams. Both require good trackers in the vicinity of the dam. Thermal cameras, daylight cameras, and laser-illuminated cameras work well for this environment.

### Pipeline and Railway

Pipelines and railroad tracks are similar in that they require long and narrow coverage zones. In many locations, the width of cleared land around pipelines and railroad tracks may only be a few meters wide in places. Power lines often run alongside many railroad tracks as well. So a fixed- beam radar is perfect for this application (see Figure 8-14). Pulsed-Doppler radar remove tree echo and flying debris in these tight areas. This is one of the few applications where fixed cameras (non-rotating) are a fine fit for the application, although spinning cameras offer some advantages of watching the departing intruder. FMCW and Pulsed-Doppler radars will work perfect for larger open areas.

*Figure 8-13. A combination thermal and CCTV camera combined with rotating, pulsed-Doppler radar protects this dam and lake.*

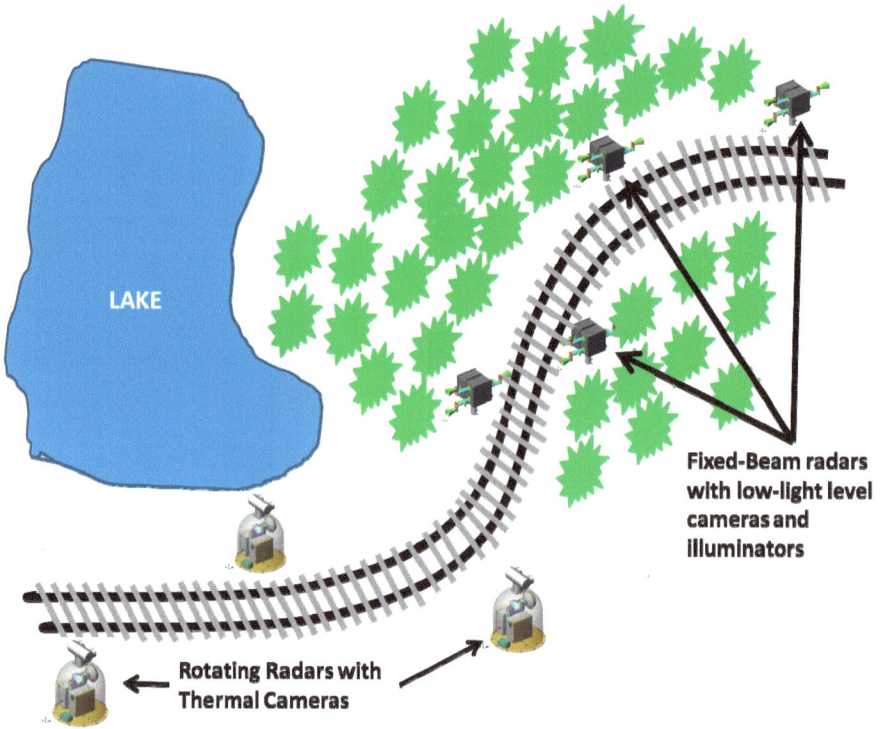

*Figure 8-14. Railroad tracks can be secured with fixed-beam radars in narrow areas or fast spinning radars is open areas.*

Figure 8-14 is an example of railroad security. In the close quarters of the forest, the fixed-beam radars offers superior update speeds. They are combined with thermal cameras. As the railroad tracks emerge from the forest, longer range, rotating radars supply expanded coverage and point cooled thermal or laser-illuminated cameras at threats. A side benefit to the security radars is the ability to detect large animals, falling trees, or vehicles on the railroad tracks.

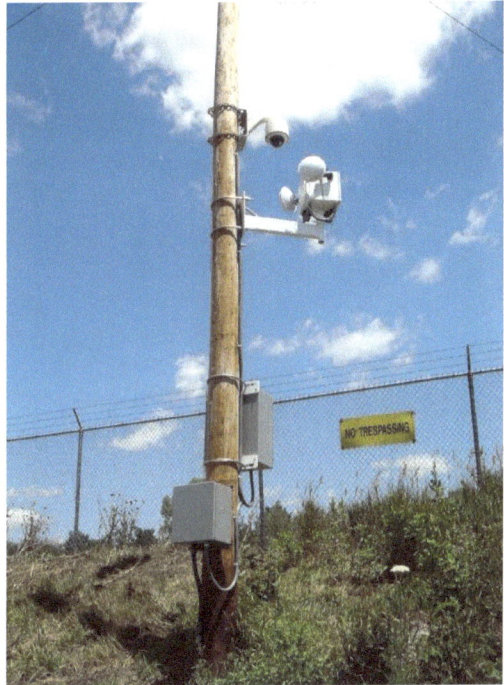

*Figure 8-15. A fixed-beam radar system offers perfect protection for railroads and pipelines. The above pictures show the versatility in terms of mounting. These radar on the left is pole-mounted and on the right it is telephone pole mounted.     Both radars are pointing low-light, security dome cameras.   Adding a low cost illuminator along each radar beam direction makes this a good performing, and cost effective solution.*

## Island Security

The security of islands requires significant planning.   Islands must be able to generate their own power or import it from underwater cables, microwaves, or tall power lines (for islands near the mainland).   Stable power is required for good security, so a substantial UPS system should be part of the security plan.     Those plans should also include communications to the mainland.   For remote islands, satellite uplinks are required.  These are expensive and do not typically have sufficient bandwidth for sustained, high-quality video.   So slower frame rates or high compression are needed. Compression software, such as ZyGoCaster, help squeeze video onto low bandwidth connections.

This author has seen the tendency of customers to put one tall tower with a single long-range radar and camera in the center of the island that must secure 360 degrees. This is done with the idea that it is lower cost and easier to maintain. Most of the time, however, this is not a good approach.

Placing a radar and camera on a tall centrally based tower on an island makes certain that the tower is going to be struck by lightening. If the area is prone to hurricanes or high winds, the tower will need to be extremely strong and may require substantial guying. In order to see the beaches, the sensors must be mounted high, so the spotlight effect mentioned earlier can occur. And the taller the tower, the more expensive it is to buid and upkeep.

A far better solution is to ring the island with towers that are no taller than the tallest trees. These do not need to be constructed in site of each other, but there should be sufficient quantity to have overlapping radar coverage in the surf zone of the island. Camera coverage should be the same as the radar. If the tower is no taller than the maximum tree heights, the tower should be able to withstand high winds much better and may not need guying.

# 9 Command Centers

## *General Information*

The sensor site surveyor will be required to also provide siting recommendations for command centers. Command centers or stations for large sites use to be like the old videos and newsreels from Kennedy Space Center in Cape Canaveral. There would be large walls of video screens, which would gingerly become like wallpaper to the sleepy guard or operator.

Today's command centers are smaller in size and much more sophisticated. Digital video and radar tracks overlaid on map images are the norm. Network video recorders (NVR's) have replaced tape decks and even digital video recorders (DVR's). Nests or rooms of wires have been replaced with network cables and fiber.

Most command centers are hosted on computer racks. These racks host the network switches, rack mounted computer or CPU blades, UPS (uninterruptable power supplies), and various other electronic equipment. The monitors have mostly moved to wall-mounted flat panel analog and digital LCD screens.    Monitors can be stitched together to form a wall of dynamic images or tiled video.

*Figure 9-1. This is a command station built by DMT.  It is assembled here for test prior to shipment.   All CPU's, video storage, and networking equipment are on the rack along with the UPS.  The larger monitors are situation awareness displays and the 3 desk monitors are operator displays.   All monitors are tied to the same desktop, so information can be slid between them effortlessly using a mouse. If the system must be moved quickly, just the rack is all that is necessary.  On the rack is a built-in sunlight-readable monitor and a series of video monitors  that can take digital and analog feeds.*

Considerable thought should be given to the number of command stations on a site.   It is not good to have only one monitoring point for a site.   It is far better, even if it is just a subset or mobile version, to have a second command station located on the premises of the site or remotely.

# *Site Surveying Concerns for Command Centers*

There are a number of concerns when siting command centers. First determine if the responding police or armed guard force is to be housed in the same location as the command center. If this is the case, then consider placing the command center within the site of the main entrance to the facility. If the forces are not co-located, then consider placing the command center at the closet point to the most secure location on the site. If the eyes of the security system go blind, then the security system becomes totally ineffective.

Unless the site is small, the security command station should not be placed at a building entrance to be monitored by the front desk person or guard. This is a critical oversight of many facilities. They spend large sums on sensors and then leave their display system vulnerable by a person at the front desk that is overwhelmed by activity.

The command center should be in a room with seating for at least 3 people, even if there is only one operator. This will enable the security chief to come in and look at logs without getting in the way of the operator. This also leaves some room for an armed response person if action is required immediately. If space permits and the air conditioning can keep up with added heat, it is better to have the computer and other command center equipment in the same room as the monitoring displays. Besides being convenient for troubleshooting and data accessibility, the operator insures the security the equipment itself. And if the system is ITAR controlled, it helps prove the system is properly protected at all times.

The site surveyor must plan for dedicated and backup power for the command center. UPS (uninterruptible power supplies) systems should have healthy battery life and should automatic sense and come on in the event of power loss.

Cable runs for communication should be planned as well. Access to the cable runs must be protected and heavily shielded. If the system has a microwave link to remote sensors, then a tower or roof mounting of the transceiver equipment will be required. Consult a communication engineer or installer to determine the height of tower needed to properly communicate with the remote sensors. Line of sight is not the only guideline for communication paths. Other factors, such as the Fresnel regions, must be considered.

If there are outside towers or mounts needed, then these should also have security. This can be physical security, such as fence around the base of the tower. A security camera should be a bare minimum add-on to any physical security measures.

The command center should have biometric security or at least passcode security at the door. Unless the glass is bullet proof, there should be no windows in the command center room. If the center in a multi-story building, it should be on the bottom floor.

# 10 Writing the Site Survey Report

## *Report Preparation*

When writing a site survey report, it is important to be concise and avoid frills and flowery writing styles. Use bullets, lists and tables to relay important facts, data, and results.

All terms should be explained in detail and jargon should be avoided. All acronyms should be spelled out and defined if uncommon.

The site surveyor needs to be able to generate a good survey report. To do so, their computer should have the following software:

- Microsoft Office, or Open Office, or similar word processing, spreadsheet, and presentation package;

- Adobe Acrobat or similar PDF file generator;

- Google Earth, ESRI Maps, or use Google Maps or Windows Live Maps.

The report should be written using a word processing and then submitted as a PDF, which will ensure the document will be printed the same way for all printers.   PowerPoint or Visio are good substitutions for a drawing package.   Make sure you have a cover sheet with title of the project, date and your name.   The second page should be a table of contents.

## *Elements of the Site Survey Report*

- **Introduction**
    - Maximum two pages in length.
    - Summarize the project, the site, and the required equipment needed for the project.
    - List all members of the site surveying team and their specialty and responsibility.
- **Deliverables (if contract has already been awarded to system integrator or sensor providers)**
    - Describe in full detail each piece of equipment that is involved in the project.
    - All equipment should be listed in a table form.  If there is more than one site in the project, but sure to create a table of equipment for each site.
- **Risk Assessment**
    - Identify the type of intruders and their likely intents.
    - Identify likely approaches to the site by intruders.
    - Describe the type of equipment that minimizes the risk.
- **Requirements**
    - Either repeat requirements if they exist, or create requirements based on risk assessment.
    - Create a table the itemizes the requirements and then show what equipment will be used to address this requirement.
    - Identify the quantity of each piece of equipment.   If you believe the contracted quantities are inadequate, carefully

explain this.  This can be a sensitive subject to the prime contractor or customer.  Sometimes it is better to recommend the extra equipment needed as potential future upgrades.

- **Site Map**
    o Site Map with Sensor Locations, Cable Runs, Command Center(s) location
    o Site Map with Radar Coverage
    o Site Map with Camera Coverage
    o Site Map with Both Radar and Camera Coverage
    o Site Map with Secondary Solution (backup plan) – Optional
- **Elevation View**
    o Show placement of sensors from directions in which they are most visible.
    o For radars, show where the vertical beam of the radar intersects the ground.
- **Performance Factors**
    o Discuss how performance is impacted by system placement.
    o List all of these assumptions used in the placement.  This is extremely important in case performance is not as anticipated.
    o Point out how sensor placement supports easy installation and maintenance.
- **Recommendations**
    o Installation Concerns
    o Action Items
    o Recommended future sensor purchases and placements and the reason one should consider these.  Tread delicately when making these recommendations and refer back to "Requirements" section.
- **Dictionary and Appendices and References**
    o List websites and references used
    o Define security and security terms
    o Appendices as necessary

- **Site Surveyor Qualifications and Certifications**
    - This is the place to give detailed information about every member of the site survey team.  Also include those that were consulted – even if they were not present for any part of the surveys.  The larger the list, the more credible the survey – especially for large projects.
    - List only those certifications and qualifications that is important to the customer.  Don't water down important accomplishments by burying them between unimportant facts.
    - Middle Eastern and overseas customers like to see copies of these certifications if they are important.
    - Recommendations are very important and should be included.   Contact the recommender to make sure they are not taken by surprise by someone calling.
- **Contact Information**
    - Encourage written correspondence, especially when it is direction or instruction to you the surveyor.   Referencing a call is often indefensible.
    - Provide email, fax, address, and office telephone information.   A professional blog or website should also be listed is available.

# Appendices

# APPENDIX A: Rules of Thumb

(1) Page 15.  The walking human is generally considered to be 1 square meter (0 dBsm).  However, as a person is walking the signature can vary greatly.  Variations in excess of 0.4 to 1.3 square meters is possible.

(2) Page 17.  An increase in power or RCS or a factor of 10 will result in a range increase of 1.778 times.  The same is true for transmitted power -- increasing it 10 times will yield a range increase of 1.778 times.

(3) Page 29.  To operate better in all weather, the security radar should have 1 to 10 GHz operating frequency.  Since antenna size is a function of frequency, frequencies of 8 to 10 GHz make for a more manageable antenna and radar size.

(4) Page 37.  When the Johnson Criteria is used by a  camera company to specify performance, use its Recognition range value to judge performance.  The customer will be less happy if the Detection range is used.

(5) Page 55.  The stand-off distance from a wall for a radar is calculated as

$$\text{Minimum distance of radar from a wall} = x * \tan(\Theta/2)$$

(6) Page 69.  2 MBPS bandwidth for communications is the minimum one should assign for video.

(7) Page 74. When installing a tower, grounding is extremely important for the survival of the sensors.  The tower ground resistance goal should be 5 ohms or less.

(8) Page 87.  When securing any four-sided facility,  three radar systems and thermal cameras is a cost effective way of providing redundant security.  Two systems may work, but offers no redundancy and will have a lower probability of detection and slower update rates.

## APPENDIX B: Useful Formulas

# Antenna Far-Field

The *far field* of an antenna is the point where the energy emitted from the radar is mostly a plane wave. At this point, the beam will be fully formed. Inside the area, the pattern grows broader in both azimuth and elevation as one nears the radar. Radar systems may need to be elevated higher than expected, or moved from near-in blockage due to the near field effect.

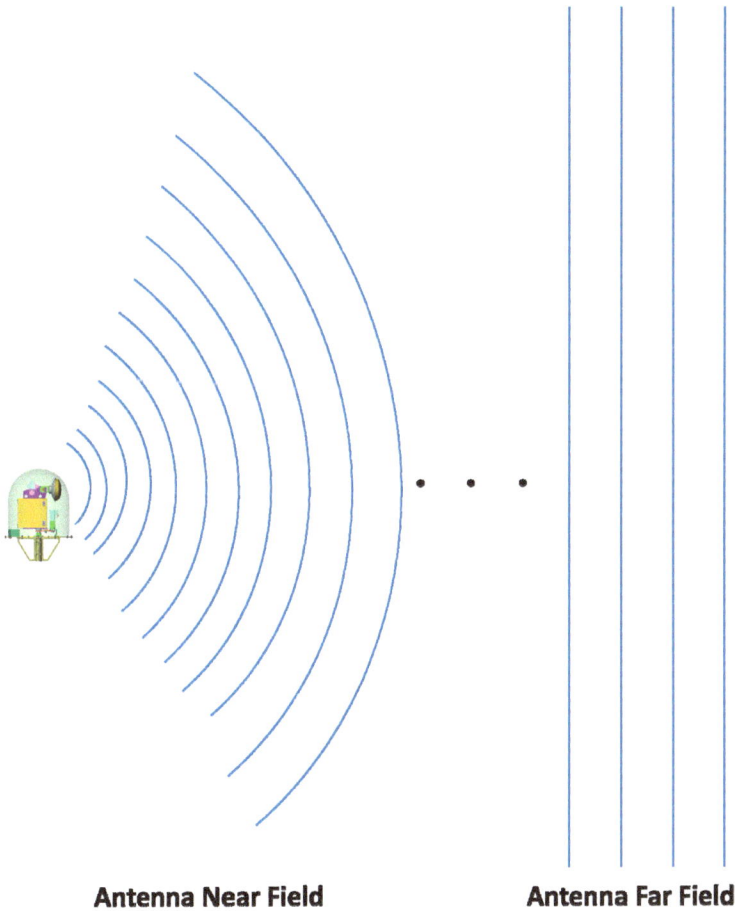

**Antenna Near Field**                **Antenna Far Field**

*Figure B-1. The near-field to the left has curved waves; whereas, the far-field on the right shows plane waves.*

To calculated the far field of an antenna, use the following formula:

$$R \text{ (range to far field)} = \frac{2D^2}{\lambda}$$

where, D = diameter of antenna

$\lambda$ = wavelength (same units as D) = speed of light / frequency (Hz)

For example, for an X-Band antenna with a 3.3 cm wavelength and 12-inch antenna yields an antenna far field of 18.47 feet (5.63 m).  Some RF engineers will use $4 D^2 / \lambda$ instead of $2D^2 / \lambda$ in the equation to be safe.

## Antenna Gain

*Gain* is a measure of the effectiveness of an antenna to focus the energy. The larger the gain, the longer the range of the radar.  Gain is calculated using the formula:

$$G = \frac{4\pi A \eta}{\lambda^2}$$

where,

A = area of the antenna in square centimeters

$\lambda$ = wavelength (centimeters) = speed of light / frequency (Hz)

$\eta$ = antenna efficiency, which is a number normally between 0.5 to 0.75

The units for gain are in dB (decibels), so take the above result and convert it to dB by: GdB = 10* Log10(G).  On the Windows calculator, just press the log button and multiply by 10.

Gain can also be calculated if you know the azimuth and elevation beamwidths for radars with parabolic (dish) antennas:

$$G_{parabolic} = \frac{160^2}{\theta_{AZ}\, \theta_{EL}}$$

Be sure to convert this to dB after computing.

# Range Accuracy

For pulsed radars, the ***range accuracy*** of a radar is a function of pulsewidth and the signal-to-noise ratio (S/N).

$$\text{Range accuracy}_{pulsed} = \frac{c\tau}{4(S/N)^{0.5}}$$

where c = speed of light = 3E8 meters

τ = pulsewidth  in seconds

S/N = 10 to 18 dB.

So the shorter the pulsewidth, the better the accuracy of the radar will be. (Of course, the shorter the pulsewidth the lower the average power, and thereby, the shorter the range.)   The greater the S/N ratio, the better the accuracy of the radar will be.

To convert from dB to linear units (which is required for this calculation), use this equation:

$$S/N_{linear} = 10^{((S/N \text{ in dB})/10)}$$

For FMCW and Pulse Compression radars, the accuracy is dependent on the bandwidth of the transmission, not the pulsewidth.

$$\text{Range accuracy}_{fmcw \text{ or comp pulse}} = \frac{c}{4\beta(S/N)^{0.5}}$$

where β = Bandwidth of the signal in Hz.

# Range Resolution

The ***range resolution*** is a measure of the ability of a radar to discern two closely spaced objects.   Normally it is measured in meters.   In radar, this in generally the point where the half-power signature of two objects intersect.   Range resolution for a pulsed radar is:

$$\Delta R = \frac{c\tau}{2}$$

where c = speed of light = 3E8 meters

τ = pulsewidth  in seconds.

For FMCW or Pulsed Compression radars, range resolution is:

$$\Delta R = \frac{c}{2\beta}$$

where β = Bandwidth of the signal in Hz.

## APPENDIX C: Definitions and Terminology

**Johnson Criteria** (page 37) The Johnson Criteria was evented many years ago by John Johnson. It provided a means of assessing the range of an infrared camera.

Infrared cameras use focal plane arrays (FPA). These arrays consist of rows of pixels that are sensitive to the infrared energy emitted by objects. The Johnson Criteria uses pixel counts to determine if an object is detected, recognized (e.g., a human), or identified (as a soldier carrying a gun, for instance).

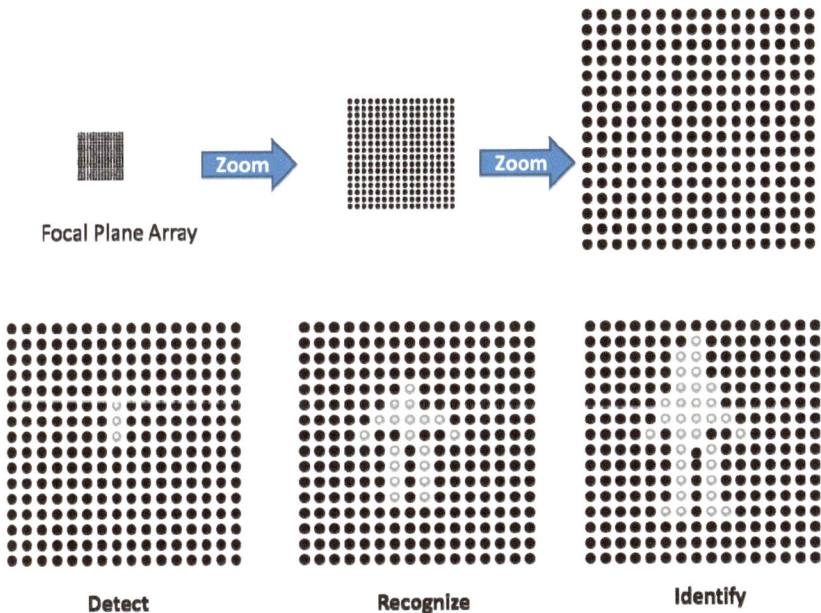

Focal Plane Array · Zoom · Zoom

Detect      Recognize      Identify

*Figure C-1. The Johnson Criteria is a criteria for determining the range capability of an infrared camera. The cartoon above is exaggerated to illustrate the Johnson Criteria. The FPA above is magnified to show the pixels.*

FLIR has a very good and downloadable link that describes the Johnson Criteria. It have very good pictures to help illustrate this using their own thermal cameras. There is also a very good discussion of performance versus the number of pixels in the array. The link is:

*http://www.flir.com/uploadedFiles/ENG_01_howfar.pdf*

119

**Main lobe** (also known as main beam, page 26):  The main lobe of an antenna is the beam of RF (radio frequency) energy that is emitted from the antenna of a radar.  Since the antenna is not infinitely wide, the energy is not completely emitted in a narrow beam.  Antennas will leak energy around the edges creating what are called sidelobes.  Radar antenna energy follows a sinx/x pattern, which is shown in the figure generated below.  This was generated using Excel.   The center large lobe would be the main lobe and the smaller lobes to each side are sidelobes.  Good antennas will have the sidelobes suppressed as much as possible.  Example antenna patterns are shown in Figures  2-5 and 2-6.

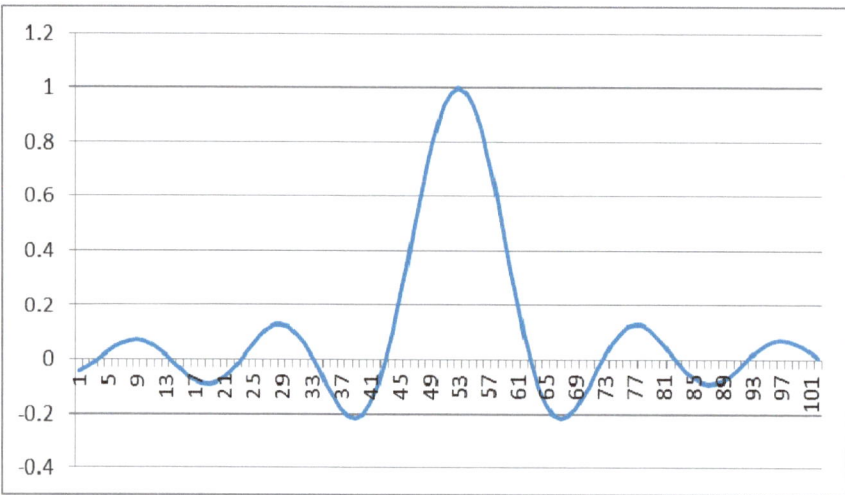

*Figure C-2.  Sin X/X function plotted using Microsoft Excel.*

**MTBF** (page 9):   Mean-Time-Between-Failure is the measure of the average time it takes for an item to break or fail.  For radar systems and cameras, this number is very important when it comes to planning for maintenance schedules and expense.    The number is best determined empirically, but is often quoted using the predictive formulas and procedures described in the US military document, MIL-HDBK-217F.

MTBF is important to the site surveyor only when it comes to planning for maintenance.    If an item has a low MTBF, it will require more maintenance.  Making sure that the planned site for the towers for these sites offer easy access for sensors that require frequent maintenance.

**Multipath** (Also known as multi-bounce, page 10): Multipath is the RF energy returned from an object that has taken a longer path than direct line of sight. The figure below shows the red line is the direct path of the RF energy; whereas, the blue line shows the multipath. The result is that the same object will show up twice on the radar screen. There can actually be many such "ghost" returns resulting from multipath.

Some radar system technologies have more problems with multipath than others. FMCW suffers from this more so that pulsed radars. So the site surveyor should take extra care to place these systems higher on towers if there are nearby scatterers (like a parking lot or fences).

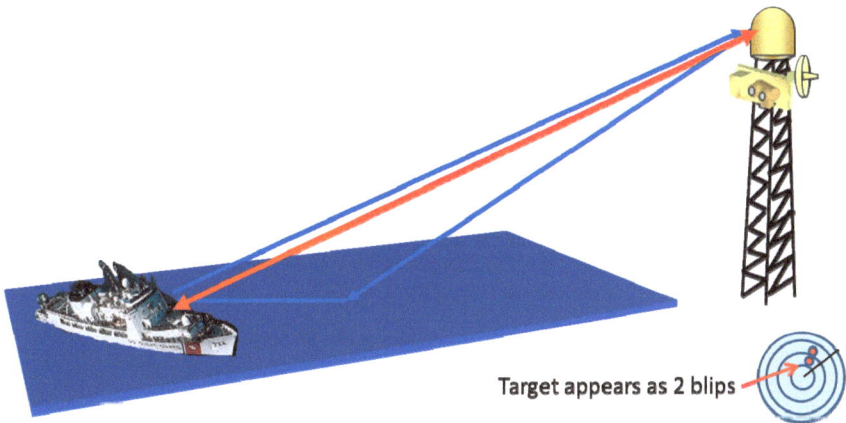

Target appears as 2 blips

*Figure C-3. Multipath illustration.*

**Waveform** (see Page 9): The waveform of a radar is basically the characterization of the emitted RF energy. The waveform for a pulsed Doppler radar, for instance, is the frequency, PRF (pulse repetition frequency), and the pulsewidth. For an FMCW radar, the waveform consists of its frequency, bandwidth, and sweep repetition frequency (i.e., how many times does it sweep over the bandwidth in one second).

**Wavelength** (Page 8) Wavelength is the distance a wave travels over 360 degrees of phase. It is best described in the figure below. Note that the wave cycle repeats itself. The number of cycles in a second is known as the frequency of the radar. 1 cycle of a wave consists of 360 degrees of phase change as it radiates through space. The length of one full cycle (before it repeats itself) is the known as the wavelength.  Wavelength is normally given in units of centimeters. The wavelength is calculated using the formula:

$$\lambda \text{ (wavelength in cm)} = \text{speed of light in cm/frequency in Hertz}$$

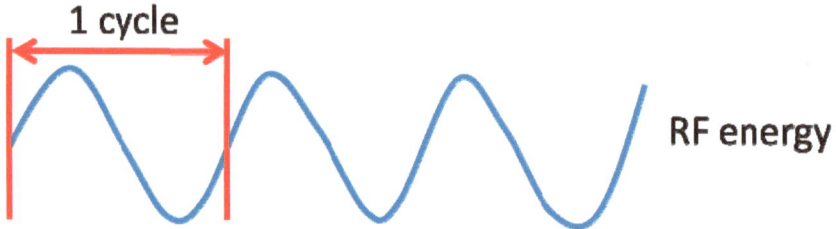

*Figure C-4. Wavelength is the length of one cycle.*

**Radomes**   Radomes are basically domes, or covers, for radar.   They protect it from the elements.   There are many kinds of radomes and they are made of many kinds of materials.   The most common radomes are made of fiberglass, plastic, and various types of polyblends.   There are very few companies that provide radomes on their radars.   DMT provides radars on all their product lines, which allows them to withstand higher winds and wider temperature ranges.   The DMT AIMS Fast-Scan radar uses Nomex radomes, which is extremely strong and is virtually transparent to the radar.

Radomes made of Nomex can support heavy weights on top, like cameras. Radomes can be infused with color or painted to match any color palette. Companies like DMT also add coatings to protect from saltwater, oils, gas, solvents and UV.  This coatings can be a flat, glossy, or shiny finish.

For the site surveyor, radomes make life easier on the site survey.   For instance, the DMT AIMS Fast-Scan can withstand tremendous winds of hurricane force.   So the radar can remain on top of towers during the storm.   Radars without storm usually have to be shut down well before the winds reach hurricane force and some require complete removal. Removal in high winds adds complexity to the tower mounts and siting.

*Figure C-5. Here are two large radomes shown in two colors. The white radome also has a glossy clear coat; whereas, the other radome has a satin finish.*

*Figure C-6. On this tower, the camera exceeds the weight limitations of the radome. The "spider" (metal legs) mount shown here supports the camera. Consult with the radar manufacturer before using this mount.*

**Tracking** (page 13): There is often confusion between tracking and detection. Detection is the first step a radar takes in alarming the operator. The radar determined from a signal that has crossed a threshold and met various criteria and declares it a detection. The detection process normally has some false alarm mitigation techniques that are applied. Once the radar has a confirmed a detection, it passes the information to a tracker.

The "tracker" is either built-in software (such as those found in the DMT radars), or third party external hardware and/or software trackers. Trackers act to link detections with similar characteristics together to form a line, or track of the intruder. Good trackers will:

- calculate position of the intruder in true Earth coordinates;
- calculate true speed of the intruder;
- calculate the course or direction the intruder is heading;
- apply smoothing or polynomial fits to the detection positions present a more representative path of the intruder;
- predict ahead the trajectory of the intruder between detection updates;
- drop those tracks that have no further updates;
- remove false alarms, because they do not endure over time in a sufficient or consistent manner;
- collapse multiple detections of a distributed object (large object with many facets or shapes) into one track update point;
- have logic that can handle maneuvering objects – not just straight line trajectories;
- handle crossing tracks of two or more objects;
- issue a unique number, name, or symbol known as the track ID.

Command center software will sometimes have tracker software. The only real benefits to a third-party tracker at the command center is the ability to lower processing load on individual radars, to do a better job of tracking if the radar does not have a good tracker, or to provide target correlation. For border applications, the target correlation is important. Target correlation is the ability:

- to take the same intruder being reported from two or more radars and recognize it is only one intruder;

- keep only one track ID number propagated across all radars the intruder may travel through.

When detections first pass into the tracker, they must be first compared to existing tracks to see if it is part of an existing track or a new track. This process is normally referred to as association. New tracks are usually passed to the track initiation process, which maintains them as unreported until there are sufficient updates to consider reporting it. This process removes many false tracks that can be started from false or nuisance detections. Doppler radars will normally have tracks reporting by the second or third detection because they have much more information (such as high precision ranging and speed) on each detection for the decision process. Most FMCW radars will need many updates before reporting because they have only position information. This author has seen vendors place FMCW radars to 20 updates before reporting tracks – especially when shooting over choppy water. In determining the coverage range for a site, the radar's time to report a track becomes an issue to consider.

There are numerous books and continuous study on tracker algorithms. The most popular today are Alpha-Beta (α-β) filter and the Kalman filter for security radars.

The Alpha-Beta filter is uses speed and position in smoothing and prediction. Acceleration makes it more robust, and is referred to then as an Alpha-Beta-Gamma (α-β-γ) filter. For ground and waterside tracking of humans and vehicles, Alpha-Beta trackers are more than adequate. Because they require less overhead and are numerically simple to calculate, they permit faster operation and larger numbers of tracks to be handled by the radar over its rival the Kalman filter tracker.

The Kalman filter is a matrix-based tracker that requires some a-priori knowledge of the intruder to be efficient and accurate. The Kalman filter can handle maneuvering targets better than the Alpha-Beta tracker. If it is properly tuned, it will home in on a more accurate trajectory of the intruder than the Alpha-Beta filter. But this is only true if the errors in position and speed are small from the radar. So for weaker signals, there is no accuracy advantage.

## Connect-The-Dots Track

Detections

Momentary reduction of radar
accuracy due to low signal
strength from intruder

## Smoothed (Fitted) Track

Predicted
Track is accurate

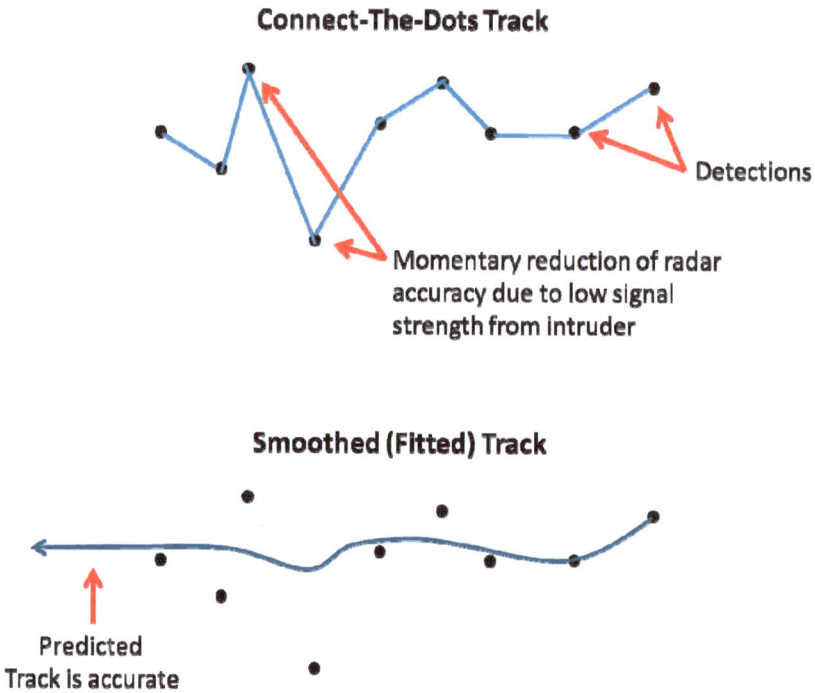

*Figure C-7. The top plot almost looks like a constellation, but it is
actually the plot of points from a set of detections. Without a true
tracker, the path of the intruder looks erratic. The bottom plot shows
the smoothed track from a modern track algorithm. The track here is
very close to the true path of the object. Its performance when the
radar accuracy is poor is one of the key reasons to use trackers.*

In the case of security radars, performance of against humans is similar
between Alpha-Beta and Kalman filters. For straight-line and typical
maneuvers of cars and boats, they are also pretty similar in performance.
For jet skis and helicopters, the Kalman filter is better if properly tuned.
But in all instances, the Alpha-Beta can track more objects and do so
faster given the same processing and memory.

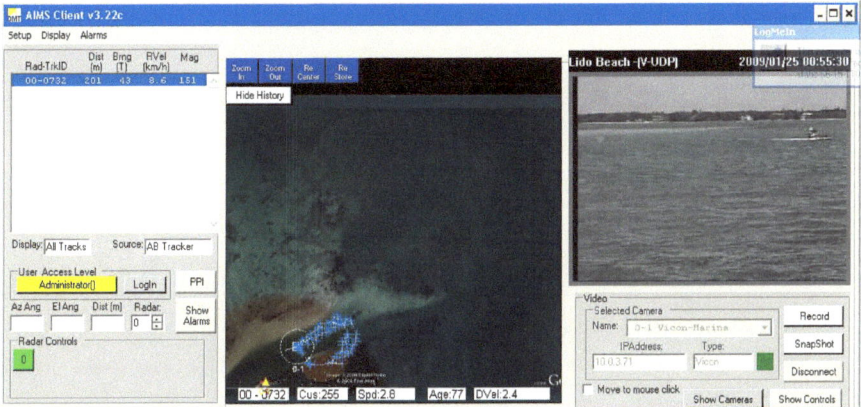

*Figure C-8. This is a command center screenshot from a radar and camera combination on a shoreline in Florida. The map display shows a satellite image of area in the center and the radar track of a small craft overlaid on it. This radar has an Alpha-Beta Tracker. Note the track is circular. The track update information is on the text box to the left and the boat is shown in the integrated video image to the right.*

*Eddie R Hughes*

# APPENDIX D:  Units

The tables found in this appendix contain conversions between commonly used units.  In addition, there are other conversions between radar and other parameters.

**Table D-1.  This table has distance units and angular conversions.**

| Item | Conversion |
|---|---|
| 1 centimeter (cm) | 0.393700787 inches |
| 1 meter (m) | 39.3700787 inches<br>3.2808399 feet |
| 1 kilometer (km) | 3,280.8399 feet<br>0.621371192 miles |
| 1 inch | 2.54 centimeters |
| 1 foot | 0.3048 meters |
| 1 mile | 1.609344 km<br>0.868976242 nm<br>5280 feet |
| 1 nautical mile (nm) | 1.85200 km<br>1.15077945 mile<br>6,076.11549 feet |
| RPM | Degrees/sec |
| 1.67 | 10 |
| 2 | 12 |
| 2.5 | 15 |
| 4 | 24 |
| 5 | 30 |
| 10 | 60 |
| 20 | 120 |
| 24 | 144 |
| 33.3 | 200 |
| 60 | 360 |

128

**Table D-2. This table has temperature conversions.**

| Centigrade | Fahrenheit |
|------------|------------|
| -40 | -40 |
| -20 | -4 |
| 0 | 32 |
| 40 | 104 |
| 50 | 122 |
| 60 | 140 |
| 70 | 158 |

Table D-3. Frequency is proportional to the inverse of wavelength. This table shows the associated wavelength of the most common security radar frequencies.

| Frequency (GHz) | Wavelength (cm) |
|-----------------|-----------------|
| 9 | 3.33 |
| 9.2 | 3.26 |
| 9.4 | 3.19 |
| 9.5 | 3.16 |
| 10 | 3.00 |
| 12 | 2.50 |
| 14 | 2.14 |
| 16 | 1.88 |
| 17 | 1.76 |
| 18 | 1.67 |
| 24 | 1.25 |

# INDEX

# Free Software with Book

This book comes with free software.  The software is an Excel spreadsheet that has most of the formulas and all of the tables contained in this book.  It is copyrighted by Eddie R Hughes and cannot be distributed without written consent.

To obtain the software, send an email to:

**register@deepseapublishing.com**

An email reply message will provided requesting proof of purchase of this book.  The spreadsheet will then be forwarded via email or a download link will be provided.

# About The Author

Eddie R Hughes lives in Herndon, Virginia with his wife and two daughters. He is the President and co-founder of Detection Monitoring Technologies (DMT, LLC), one of the world's premier builders of radar systems. Mr. Hughes is an author and has written both technical references as well as novels, including the series of books known as The Bryant Family Chronicles.

For over 28 years, Mr. Hughes has worked in radar for companies such as XonTech, Inc., Lockheed, Sperry Marine, and DMT. He has served as radar subject matter expert for National Missile Defense. Mr. Hughes designed and developed the XonTrak radar system for XonTech, and the AIMS Fast-Scan and the AIMS Fixed-Beam radar systems for DMT. Mr. Hughes was the lead engineer for the development of the first production impulse radar back in the late 1980's for Sperry. He redesigned a more cost effective solution for Sperry's IDAS radar and designed and built the first high-gain ultrawideband antennas for it.

*Author Eddie Hughes*

Mr. Hughes developed all the initial algorithms and software used in the XonTrak radars, in the IDAS radar, in the AIMS Fast-Scan and AIMS Fixed-Beam radars. He also developed advanced RCS software for the US Navy, and the National Missile Defense.

Mr. Hughes has traveled the world to install, support, and sell integrated radar and camera products. These products have been installed and used in places like Norway, France, Italy, Iraq, Saudi Arabia, the UAE and in the USA to mention a few.   Mr. Hughes has performed site surveys, helped with installation, and supported all the radars in these locations.

*Eddie Hughes supporting operations in Iraq.*

*Eddie Hughes going up to help on radar install on the world's largest yacht.*

133

# Upcoming Books and Information

Interested in more technical references, training programs, science- based fictional work, or software?   Visit the Deep Sea Publishing website:

## www.deepseapublishing.com.

The website provides more information about the author, book signing events, downloads of radar spreadsheets, and more about the technology found in the books. You can also leave messages for the author and others on the blog.

Deep Sea Publishing (DSP) is a Florida-based company that sells fictional novels and technical references. The website supplies details all DSP publications and the expected release dates on new material.

Deep Sea Publishing books may be purchased in electronic or paper formats. Check the DSP website for a list of resellers.

www.ingramcontent.com/pod-product-compliance
Lightning Source LLC
Chambersburg PA
CBHW041220270326
41932CB00003B/12